T0258513

THE HIDDEN GEOMETRY
OF FLOWERS

THE HIDDEN GEOMETRY OF FLOWERS

Living Rhythms, Form and Number

KEITH CRITCHLOW

Floris Books

First published in 2011 by Floris Books
Fifth printing 2022
© 2011 Keith Critchlow

All photographs and geometric line drawings are by Keith Critchlow
unless otherwise stated in the Image Credits
Colour photographs and geometric line drawings © 2011 Keith Critchlow

British Library CIP Data available
ISBN 978-086315-806-3
Printed in Great Britain by Bell & Bain, Ltd

 Floris Books supports sustainable forest management
by printing this book on materials made from wood that
comes from responsible sources and reclaimed material

MIX
Paper from
responsible sources
FSC® C007785
FSC
www.fsc.org

Contents

Part Three: The Geometry of Flowers

Part Four: Flowers of Geometry

Part Five: Leaves and Life's Most Consistent Miracle: Photosynthesis

Part Six: Conclusions

CLARENCE HOUSE

Our scientific study of the outward mechanisms of Nature has no doubt been the springboard for the tremendous technological advances that have transformed the lives of millions of people, but there is much to discover – or rather "recover" – if we observe Nature as Keith Critchlow has done to create this great labour of love. It is a sad fact that our modern outlook does not recognize geometry as a language by which we may understand Divine order. Keith is one of the very few who has swum against the tide to keep this flame alive and here he sums up his life's work, essentially by posing a very important question. If it is the case, as we are persuaded to believe today, that Nature is little more than a utilitarian collection of mechanical processes, why should a flower be beautiful? After all, Nature is so remarkably efficient; she does not do anything without good cause. Keith's answer is that the "good cause" is quite simple. The beauty expressed by a flower's hidden geometry is as necessary to the world as its reproductive function is to the plant. How fascinating, though, that this beauty is communicated not by the plant, but by us. It is we who respond to its beauty. It is humanity that has always embedded this geometry in the world's greatest works of sacred art and architecture, simply because we resonate with these hidden patterns. We, too, are made up of them and, thus, we contain the universe that contains us. Or, as the traditional philosophy for which Keith speaks would phrase it, we are each a microcosm of the macrocosm. The patterning hidden in a flower also enables all life to achieve an active state of balance that we call "harmony," which is the prerequisite of the health of each of the Earth's vital life-support systems.

Keith quotes Silesius at one point, that "God is in all things as unity is in all number." From this comes the conclusion, surely, that no one organism can grow well and true without it also being mindful and enhancing the well-being of the whole. This is the golden rule of life, expressed through a precise, naturally occurring geometry that reveals the sacred "ratio," our proper relationship with the whole of creation. It is this integrated relationship that we must become more conscious of if we hope to "re-mind" our collective view and avert the worst of the disasters that now hover darkly upon our horizon. Through his elegant study of flowers, Keith stresses what is key to the approach we have to take in the treacherous years ahead.

Acknowledgments and Gratitude

To His Royal Highness the Prince of Wales for his generous Foreword and many years of continuing support for this work. Particularly in offering the use of his garden at Highgrove as inspiration.

Acknowledgment is fundamental for any author. As nobody is an island those of us who take on the responsibility of being 'authors' have to thank all who contributed to the final book. Gratitude is the basis of awareness of the whole and eventually finds its way to the supreme source – however modest.

First must come my wife Gail who is more responsible for this book than she really knows. It is she who is the hidden wisdom and who has tended our small garden in London for over 40 years. Next come my family all of whom garden and encourage their children to be gardeners. They are Louise, Amanda, Matthew and Amelia together with all their children and even our great-grandchildren. Never forgetting my gardening father, my mother and brother, architect David Critchlow, who struggled to teach me maths many, many years ago.

Next come those who have inspired me over the years: A. S. Neill, Dr Ida Seymour (who first introduced me to Plato), Richard Buckminster Fuller, Dr Kathleen Raine, Frithjof Schuon, Titus Burkhardt, Warren and Rebecca Kenton, Dr Rupert and Jill Sheldrake, John Michel, Brian Keeble, Sir George Trevelyan, Dr Wendell and Tanya Berry. William I. Thompson and the Lindisfarne Group, The Very Rev. James Morton, James Lovelock and Lama Govinda, S. H. Nasr, John and Nancy Todd (The Ark Originators), Thomas Banyacya Jr., Helen Whittaker, Keith Barley, Isaac Tigrett, Carla Marchesan, Trisha Mulholland, Sathya Sai Baba, Sri Venkatesh and family. Dr Raj Patel and family. For special literary help, thank you to Ian Skelly of Temenos and Glenn Storhaug of Five Seasons Press. Next Dr Julian Henriques and family for their enthusiasm and special interest. Thank you to all members and contributors of the Kairos Foundation, and to Amanda for managing Kairos for so long.

Next come my students, colleagues and ex-students at the Prince's School of Traditional Arts. Paul Marchant, Dr Khaled Azzam, Dr Emily Pott, Emma Clark, Simon Trethewey, Jonathan Horning and Ririko Suzuki. After this my long standing friend and sustainer of the Temenos Academy, Stephen Overy and his family, not forgetting Vinod Taylor,

Sir Nicholas Pearson, Prof. David Cadman et al., and the august council; Gail Thomas, Robert Sardello and all the Dallas Institute, Darlene and David Yarbrough, Hanne & Maurice Strong, Thomas Neurath, John Harrison & John Allen.

Next thanks is especially due to Ana Maria Giraldo for her dedicated contribution to all aspects of this book, Graham Challifour for his much valued assistance and support, Marsha Andreola for her geometric work and her enthusiasm and Tom Bree for his research. Further thanks go to those with special gardens, they are: Oliver and Barbara Clauson, Nicholas and Gay Smith, Glen and Carol Baxter, Annie and Greg Harris, Vicki Hallam for introducing me to the Passion Flower. Jenny & Heywood Hill, Robin & Jessica Sutcliffe. Najeeb, Moneeb and Monika at the Creative Print Shop as well as to 'Tiger Lily', my local flower shop who made the Rose 'crown'. Finally Julian, Graham and Sandra for their exceptional support as The Twelve Healers Trust.

Last but not least the patient staff of Floris books: Christopher Moore, Catherine McKinney, Helena Waldron, Katy Lockwood-Holmes and Christian Maclean. A special thanks goes to the The Twelve Healers Trust for their generous help in securing the extra pictures. Finally gratitude and thanks to all those who have 'processed' the words and images: they are Amelia Critchlow, Somaya Critchlow and Moaki Critchlow-Castro.

Author's Preface

This book like the flowers themselves speaks primarily in the language of images. It also follows a four-layered structure. These can be called points of view. The first looks into the tangible structure of flowers, the second takes account of the social value flowers have for us. The third concerns the symbolic or cultural use of flowers. The fourth celebrates the inspirational effect flowers have on us. All four are integral as well as existing within their own separate contexts.

This is not an 'easy read' book that follows a single flow of reasoning from start to finish. On the contrary it is composed of insights as well as outsights, focusing on how we regard flowers. It is designed to encourage all who read it to look at flowers in a new way. There are also pauses, during which the reader is encouraged to turn to the nearest flower and contemplate it and hopefully see it anew.

The illustrations were hand-drawn by the author. Geometry can be considered from at least three viewpoints. First as a technical exercise mostly serving industrialization. Secondly as a purely mathematical function. Thirdly, and most importantly, as a science of the soul. This has to be performed with the human hand and is fundamental to a deeper understanding of the Platonic wisdom tradition. Geometry is only fully understood by doing it.

None of the ideas contained here is dogmatic or fixed, but rather an offering for consideration. We have been guided ourselves by the truth of flowers, their beauty and what makes them so important to us — maybe they are also our teachers of the time-honoured objective truths of number, geometry, harmony and wholeness.

Keith Critchlow
London, December 2010

Exile

Then, I had no doubt
That snowdrops, violets, all creatures, I myself
Were lovely, were loved, were love.
Look, they said,
And I had only to look deep into the heart,
Dark, deep into the violet, and there read,
Before I knew of any word for flower or love,
The flower, the love, the word.

Kathleen Raine

The Blue Flower

You are the pilgrim's friend,
Moment of bright beauty
Along the hard road. You're kind,
Offering a restful harmony
That's served by sun and ground and air.
Yet still we cannot tell
Your silent purpose here.
Your fleeting passage of a day
Remains a mystery. We'll find
Your meaning at the hard road's end.

C. J. Moore

ONE

INTRODUCTION

Life sends up in blades of grass
Its silent hymn of praise
To the unnamed light.

Rabindranath Tagore

How clearly an open flower like this Marguerite displays
its 'flatness' to the sky and sun.

Some Talk of a Return to Nature

Everyone has to seek nature for himself.

Masanobu Fukuoka

Life is the most sacred thing in existence.

J. Krishnamurti

Life sends up in blades of grass
Its silent hymn of praise
To the unnamed light.

Rabindranath Tagore

The experience of a radically desacralized nature is a recent
phenomenon, moreover, it is an experience accessible only to
a minority in modern societies, especially scientists.

Mircea Eliade

Chartres Cathedral from the edge of the city. The flowers, the trees, the produce of the land are all part and parcel of ourselves. We build in gratitude for the gift of nature of which we are an inseparable part.

It is true, some do talk of a return to Nature — and I wonder where they have been and what they think they are. It is more than apparent that we urgently need to review and change our attitude and behaviour towards the material world. It is as if we had buried our heads deeply into the superficial and completely distracting worlds of information technology, the luxury industries, and the so-called news media of the 'papers', television and radio. Our integral partnership with the natural world has been so neglected that children no longer learn that the milk we drink comes from cows or goats, or that the processed cereals we eat come directly from grasses grown in Mother Earth. Even farming has become 'agri-business' when it clearly needs to be *agri-culture.* Education has become 'edubusiness' when it clearly needs to be *edu-care.* Medicine and health have become the domain of the powerful drug companies, when the sources of our most commonly used medicines have their origins in the herbs and plants of the natural world.

Apples 'close pack' like any spherical forms because the blossoms are so naturally close.

The importance of the bee cannot be overemphasized. Flowers (in this photograph, Echinacea) naturally offer their lives to the insect world in a symbiosis.

We know that the statement 'All flesh is grass' (Isaiah 40:6) is as much a scientific fact as a biblical truth. This means that all of our body is fuelled by vegetation. In our hurry to chase false and distracting values we too easily forget the necessity of recognizing which flowers, fruits and plants are beneficial to us and which are poisonous. Even the definition of poison has become obscured through multiple processing and the efforts of the food industry to disguise exactly what it is we are being offered to eat. I wonder what proportion of processed food is actually 'poisonous'?

This introductory book is a commitment by a lover of flowers, not a scientist (in the modern usage of the term), to offer a new perspective on our relationship with the natural world. In the view of the present author, there is nothing within this offering that has not been common wisdom in past ages; rather it is an attempt to re-engage the human mind with its overall interrelatedness to the rest of the world. We have heard of the movement of the butterfly's wings that can start a hurricane, so even the most modest questioning of human values can produce far-reaching changes in our social wellbeing. Here we have chosen to review the role of flowers in the whole scheme of things.

Flowers are quite evidently vastly important members of the fully-functioning whole we call Nature or Life. However, we have to be careful not to limit them to their obvious material or mechanical function of being the fertility mechanism of the vegetal world. Flowers are good, flowers are beautiful. Flowers are true to their kind and to their species as the consistent providers of the fruits and seeds of each plant. And not least they give joy to animals and insects, and to ourselves.

In this study of flowers and their geometry we are going to use a time-honoured approach for viewing the subject. This is to propose that there are four different positions or perspectives for viewing any subject. The first is the *evident or material level* — this is the bodily plant or flower. The second is the *social or psychological level* or how flowers affect us as people both actually and symbolically. The third is the *cultural and mythological level;* this is the deeper cultural significance of flowers and the plant world, which differs according to each culture and civilization. The fourth we choose to call the *inspirational level* — that which moves or inspires all human life. This final level is the motivating energy of all the rest. Each level invites participants 'up' to a more subtle level of 'seeing', 'understanding', and state of being. Each level offers its own panorama of vision, and each offers its aspects of 'permanence' according to the ability of the receiver to accept what is proffered.

Flowers speak to us in their own language, that of beauty, and across all levels. How many of us have been halted in our tracks by coming across a modest, yet beautifully coloured and perfectly symmetrical, wild flower while walking on a cliff top overlooking the vast ocean? It is just this extraordinary vulnerability of a flower that is so disarming. It cannot run away. It attracts the attention of the insects which will assist in its fertilization and the attention of the humans or animals which 'recognize' it. The flower offers many levels of value to the finder, from delight in finding beauty in its colour, fragrance and symmetry to its medicinal benefits. In this book we have chosen to focus on an aspect of flowers that has received perhaps the least attention. This is the flower as teacher of symmetry and geometry (the 'eternal verities', as Plato called them). However, the word 'teacher' needs to be qualified as the author finds that true teaching means 'being reminded', rather than learning for the first time. This is more in line with the conclusions of the traditional wisdom of mankind. Plato talked of 'anamnesis', which was a sort of deep memory. This was based on Socrates' proposal that we have drunk of the waters of forgetfulness when we come into our incarnate state at birth, whereas previously as 'souls' we had access to 'all knowledge'. So, from birth onwards we relearn and are reminded of our intrinsic inner knowledge. Socrates taught his pupils by asking questions to draw out this knowledge: 'educe', meaning to 'lead out', being the root of the word 'education'. Other traditions use similar terms which have similar associations with 'remembering'. To re-member literally means to put the members back into their natural wholeness. It is in this sense that flowers can be treated as sources of remembering — a way of recalling our own wholeness and that of the natural world — as well as learning about our inner power of recognition and consciousness.

What is remembered? If we investigate the theory that Socrates put forward, we have a series of evident aspects of flowers to refer to. Firstly, and most obviously, they stay 'planted'. In common with all plants they tend largely to stay put. This means that the soil where they have germinated will remain their home environment throughout their life. This will be as true for the great Oak tree as for the humble Stinging Nettle. Next the flower, being the celebratory climax of the necessity of the plant to reproduce, becomes the most attention-seeking aspect of each plant. Although this is not completely true, as the leaves or bracts of some plants are clearly their most decorative element, for example Bougainvillaea where the bracts provide the bright colours, not the flowers. Each flower has a face that it puts to the world, whether earth-facing

Humanity has treated herbs for what they are worth for centuries. The knowing eye can tell the healing plants. All human societies have used healing plants and herbs.

Attention seeking as well as looking heavenward. But the pollen is vulnerable to showers of rain. Here the Poppy opens to blue sky.

The naturally 'shy' and pollen-protecting poise of these Bleeding Heart flowers.

as with the Snowdrop, or sun-following as with the great Sunflower and so many others. All members of the animal kingdom, including humans, are normally aware of the flowering domain of the plant kingdom. It is the faces of certain flowers that this book has focused on to bring out the qualities that can validly be described as 'reminders', vehicles of 'recall' and 'remembrance'. These are qualities that awaken the mind to nature's inner secrets or the intrinsic 'ideas' as Socrates might put it. These ideas are 'permanent' principles, similar to the truth of mathematics.

It is a reasonable generalization that all things in the created world are subject to conditions of their existence. The conditions that govern the growth of any plant are similar to the conditions that govern anything in existence. It is useful to remember that this word *ex*istence etymologically means 'to stand out' and to be externalized, expressed in outward acts or appearances. To exist means to 'have being in a specified place or under specified conditions' according to the *Oxford English Dictionary*. These conditions (for *all* existent things, flowers included) then offer themselves for us to consider. These are the permanent principles that underlie all experiences: firstly *space* (the theatre of events), secondly

time (the duration of events), thirdly *number* (the sequence of events), fourthly *form* (the assemblage of events) and finally, as a summarizing cohesion of the rest, *substance* (the substantiality of events). This last word 'substance' has a fascinating inner tension which is not hidden but intrinsic. Is 'substance' a reality as the materialists would wish, or an illusion as the Vedanta or Buddhism would posit? The meeting ground is at the core of consciousness itself. 'Substance' literally means *standing under* or *under-standing*. Either way it is a product of consciousness. Some prefer to call the *experience* of substance *act*uality as Einstein implied. Others refer to the consciousness of substance as based in reality, rather than actuality. ('Actuality' means the 'act' of standing out. 'Reality' can mean both the hidden timeless principle as well as the integral whole.)

In this sense, 'reality' is the *permanent* domain of pure principle or the intelligible realm, the source of consciousness itself. 'Actuality' is the way in which things act upon our consciousness, either through the senses or due to the internal reality of the imaginal world. This word 'imaginal' was used by Henri Corbin, the Sufi authority who studied the deeper meanings of the Islamic philosophers such as Ibn Arabi, Avicenna

Bell-like and rhythmic hanging flowers of Solomon's Seal.

and Suhrawadi.[1] Either way, 'substantiality' has the same foundation in consciousness or the very structure of the intelligible. Sense is one thing, making sense another.

If we put the conditions to the test a flower answers most helpfully. It requires *space* to exist. This space may be even as large as miles when it comes to pollen on the wind. It requires *time* to fulfil its life cycle and produce the wherewithal to further the species. It requires *number* as a method of differentiation and balance, from its cells to its leaves and its petals, as well as for nourishment. The flower also requires *form* in which to cohere, to be recognizable, and to fulfil its necessary biological functions — and therefore become substantial. Each flower, once it is in material existence, exhibits its actuality and reflects its reality or the archetypal pattern by which it is recognized. Each of these conditions of existence is co-dependent on being within the field of consciousness — with us as observers bearing witness.

This study will concentrate on the spatial and numeric conditions that are true for all things. In so doing, the other phenomena of existence will automatically align with the qualities we find in flowers.

Consequently, what is evident in the geometry of the face of a flower can remind us of the geometry that underlies all existence. Studying the geometry of flowers is therefore a powerful way to reconnect us with the idea that we are all one; a definition that is captured in the word *universe*.

As Heraclitus said, 'Nature loves to hide', which could also be interpreted as, 'Nature loves to hide its principles of existence'. This again is a reminder of how Plato's character Timaeus begins his explanation of the nature of the 'made' universe. 'It is fabricated', he says, 'of Sameness, Otherness, and Being (or Essence)'.[2] Such is true of all principles or laws of Nature. They are the same in principle, but expressed differently in each instance. Flowers too are recognizably of a particular plant, as are the leaves of that plant. However, each flower and each leaf is quite demonstrably different from every other leaf of the same genus or species. Analysing flowers for their archetypal geometry requires a careful and consistent discipline. This discipline is a balance between 'intuitive thinking' and a 'reasonable proposal' or the *most* likely story as Plato insists in the *Timaeus*.[3]

The face of a flower presents us not only with colour harmonies, colour patterns, fragrance and delicacies of form but also, in many cases, powerful examples of symmetry and the geometry of each part of its symmetry. In the attempt to find the nearest to an archetype, one has to take the best and most convincing specimen petal to see how it would be arranged in the natural

The hexagonal Daffodil is clearly joined at the back of the petals.

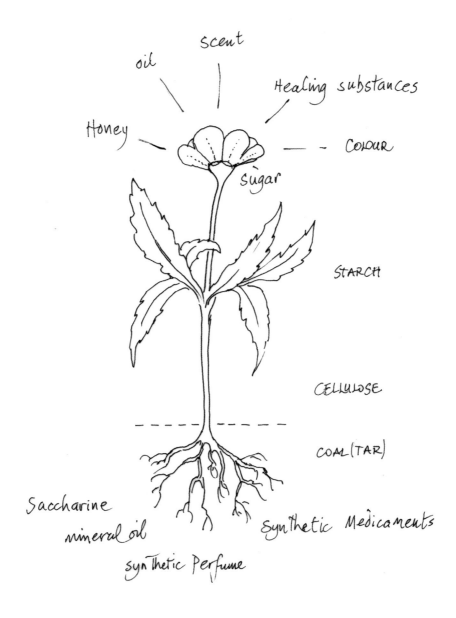

The anthroposophical view of all the uses of a plant.

Dramatic images of the spiral principle at galactic and subatomic scale.

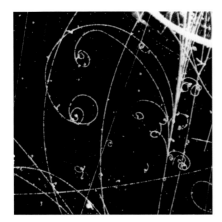

symmetry of the flower. We must judiciously bear in mind sameness, difference and wholeness, as discussed previously. How and why a particular petal and the composite flower exactly occur has to be acknowledged as a mystery. Similarly, a 'law' of nature is expressed slightly differently on each occasion. Yet both are embraced in being.

There is another factor that needs to be acknowledged: each flower is 'in process' at all times. This means that it usually spirals into being, maintains a full-faced 'aspect' for a time, then proceeds to withdraw life from this form as it contracts and often spirals back down into the soil. It is the full-faced or flat, planar face of the flower that we are studying here. Even so we are quite aware that however 'flat' the flower's face is, there will inevitably be a three-dimensional aspect that cannot be taken fully into account. This, however, points to another philosophical consideration, namely the 'dimensions' of space normally expressed as length, breadth and thickness. The two-dimensional, which is the way the human eye takes in its images, only has breadth and length, one may say. We move our heads so our eyes can scan the third dimension of its volume. So, strictly speaking, the flat face of a flower offers to us a two-dimensional object of contemplation. This naturally does not exclude the obvious experience of spherical flowers such as some forms of Dahlia or the Dandelion seedhead. What we find when investigating the geometry of the life cycle of a flower is that it describes or creates a 'flow' from a point-seed to flower bud, and then through a spherical unfolding. The field of action that the flower's movement takes up is largely spherical. Yet petals perform a winding up to a winding down in each flower to complete its life's movement. This can best be illustrated graphically as it is difficult for the reader to visualize imaginatively.

The flower world presents its use of the golden sequence in right and left-handed spirals. Here mostly 13:8 in the golden sequence.

Above. Spherical outcome of petal proliferation. Human cultivars often fill out their blossoms with more and more petals. This does not help the pollen factor in generation or fertilization.

Left. The Dandelion is an assertive flower and its seed-distributing pattern is wondrous. Another beautiful example of sphericality.

Where each seed came from. The Dandelion head displays its strategy.

What a display of colour, form, number and radiance!
This Passion Flower (*Passiflora*) is ten in its outer form.

How Do We See a Flower?

That rose with your earthly eyes you see
Has flowered in God from all eternity.

Angelus Silesius

Nature is rich and beautiful, and literally provides human indigence with
her untold wealth with the result that the properties of things overflow
into words.

John of Salisbury, from the Metalogicon Book I, Chapter 16

It would not be fully reasonable or even practical to ask somebody what effect a flower has on him or her. Each individual is likely to give a different answer, although the answers would tend to fall in groups. The personal awareness and life experiences of each viewer, even those with an educated knowledge of flowers, will inevitably influence the reply. However, the majority would probably plead that the full effect of a flower is too difficult to explain in words — as with beauty itself. Words have their limits too, as all of us know when we are overwhelmed by emotion.

All flowers offer us a huge range of possible responses. Perhaps it is the colour, perhaps the pattern of the petals, and for the knowledgeable it could be a response to the name and type of flower. It could be a desire to respond to the perfume of the flower. It could also be the group effect of flowers arranged in clusters. Another response would be to estimate what stage of the flower's unfolding has taken place and what more is to come. Nostalgia must not be forgotten and even the memories of when this type of flower was last seen — or even as a reminder of, or reference to, a favourite person, author, poet or playwright (Goethe, for example).[4]

A flower has a number of exceptional properties and it is too simple to describe it as merely the reproductive organ of the plant. We can see the flower as a celebration of the fulfilment of a plant. For if Goethe's perceptions are to be given value, as this author fervently believes, then the *whole* of a plant needs to be recognized both in space and in life cycle, let alone its special properties.

The release of the petals from the sepals is inevitably dramatic and begins a spiral dance that is profound and lasts throughout the life of the Rose.

Earthward-hanging flowers of the Fuchsia protect their pollen and display a fourfold symmetry significantly.

Some theorists have argued that only messages which are intentionally sent and received should be considered as communication. Others suggest that any behaviour or symbol that has meaning to a receiver should be regarded as communication. On this basis, all natural things that we experience communicate in some fashion, and it seems that a plant mostly communicates through its flowers. This word 'communication' needs the broadest meaning, since we are not necessarily confining it to words and logic or even to the senses of sound or touch. Communication, in our current usage, means communicating aspects of awareness or consciousness. The dictionary defines the word 'communicate' principally as to 'share' or 'transmit' information (maybe better defined as knowledge). This transmission can then be qualified by 'speech or writing', but a further meaning is present which is to do with space, such as in rooms. This is the spatial awareness of, for example, 'two spaces having a common door or opening'.[5]

If we take 'transmit' as the broadest, and in a sense the most abstract meaning of 'communication', then we are in a realm which can go beyond words or even numbers. The transmission of colour in itself is a sufficient mystery. We can talk of colour as wavelengths of light which are absorbed by, and emanate from, the surface we see as coloured — but what the *experience* of colour does for us is quite above and beyond physics. What a flower transmits to a human being can only be accurately described as multivalent or profoundly deep *and* broad.

We have already said that a flower has a 'face'. This face is usually turned towards the light — particularly the sun — but in some cases the flower faces downward like the Snowdrop or Fuchsia or Pulsatilla, no doubt protecting its delicate interior. Yet even this simple explanation is unlikely to be totally sufficient.

Flowers that raise their faces towards the light are those that communicate most directly with us (also the fragrances are most easily experienced when the flower faces upward). This enables us to recognize the symmetry, patterns and colouring of the petals, and their intrinsic geometrical proportions (although these are taken in, like most proportioning values, subconsciously or instinctively in the first instance). Only the trained eye will pick up immediately such qualities as the exact symmetry.

In short, flowers have fed the human imagination in a multitude of ways and for us to focus here on the geometry of flowers is merely one of these, but one which has received little attention beyond discussions of symmetry. Geometry is *an essential condition of existence* and so has the value of connecting flowers to all other natural objects — representing as it does the universal order of space. This unfolds inevitably from a point to a line, to a plane, to a solid, as we will often repeat.

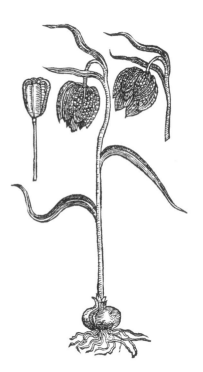

This early woodcut demonstrates acute observation.

Snake's Head Fritillary. A shy and beautiful flower named after the pattern of the checker board or the Latin term *fritillus* for 'dice box'.

The ornamental current or Ribes has an early spring flower which
signals the warmer weather which bees need before they can fly.

What is Life?

Behold my beauty, witness of me in every man,
Like the water flowing through
The sap branches.
One water drink they, yet they flower
In many hues.

Ali Ash-Shushtari

To view a flower only as a mechanical response to a series of environmental factors represents a poverty of sensibility. The fact that contemporary mechanical materialist science cannot explain what life *is,* only what it does, serves as a warning to the present author. We should be extremely wary of adopting any mechanical (or so-called functional) theory to cover our ignorance when considering as important a phenomenon as life itself. From a reductionist perspective, life is neither fully explainable nor analysable. Its permanent mystery has always been the prime motivation for human scientific endeavour of any kind. It is good to remember that all scientific investigation relies on mystery, something to be solved, something to be researched.

Life on our mother planet is dramatically, even gloriously, prolific in form, scale, number, type, aspect and, most importantly, interrelatedness. Essentially *life is one,* however much some like to focus on the aspect of local conflict within the overall biosphere. Life also nourishes life.

What is most distressing is that without a definition for life, and without a traditional reverence for life, the contemporary scientistic community gains so much of its knowledge of living creatures by killing them. ('Scientistic' is preferable to 'scientific' as it specifically refers, in the modern world, to those who conduct their work on the premise of mechanical materialism.) This is not said with any dramatic intent, rather as a cold and sobering truth. Dissecting living things into smaller and smaller parts has nothing to do with the traditional meaning of 'inner'. Life is a motive force within, and is equally a whole.

It was Fritz Schumacher[6] who described the ascent of life and consciousness as a ladder of miracles when he reminded us that the transformation from a mineral to a plant is a miraculous step. This is followed by an equally miraculous step from a plant to an animal. Not stopping here, although we are part of the animal kingdom, the further miraculous step is for us to become human (even if we must qualify this by saying 'potentially'). This means the added dimension of language, culture and, we have to add, revelation. Schumacher was particularly sensitive to the fact that each of these transformations is unquestionably miraculous, in the correct meaning of the term. In fact, the almost elementary emphasis in current usage of the term 'evolution', which is based solely on changes in physical characteristics, has virtually no value in comparison with the truly miraculous evolution from mineral to plant, plant to animal, and animal to human, nor can the movement or evolving of life explain it. Each and all of these transformations remain mutually sustaining in principle and in fact. Life flows through them all, drawing them up to the light.

To conclude this observation we would like to recommend *A New Science of Life* by Rupert Sheldrake as the best contemporary scientific summary of the subject.[7]

> Very little is actually known or even can be known
> about the details of evolution in the past. Nor is
> evolution readily observable in the present. Even on
> a timescale measured in millions of years, the origin
> of new species is rare and the origin of genera,
> families and orders rarer still.
>
> Rupert Sheldrake

Modern Science, Truth and Beauty

To 'save' a human life can only be seen as a virtue with very rare exceptions. The quality of that life after it has been saved is another consideration. We cannot but show our extreme gratitude to modern medical science, from blood analysis to dentistry, from the relief of pain to immunization and immunity support. However, the deeper issues that emerge are those that affect the quality, morality and purpose of life as a whole — not just momentary pain relief. It is very often the case that the shock and pain of an illness can deeply awaken a person to reconsider what his or her life is, or ought to be about. Thus there are always two sides to every 'new remedy'. Pain has also been called an important teacher.

The word 'science' and its use is what is most important here. There was traditionally only one essential meaning before mind was divided from matter, and soul from bodily investigation — and that was the 'science of the soul' or the art of life as a whole. There was an acceptance that bodily, psychological, cultural and inspirational health were inseparable. They were all seen as the comprehensive unity of a useful and purposeful life. This author's long-standing friend Nicholas Woodward-Smith was in charge of the family unit of each patient suffering from cancer at one of London's largest hospitals. He helped the counselling fraternity realize that it was *the whole family* that had been and may continue to suffer — so all needed consideration and help. This could be called wise common sense. An essential realization of wholeness.

A beautiful specimen flower. The power of colour, pattern and symmetry.

Brian Goodwin, biologist, mathematician and 'holistic scientist', has stated succinctly that there were three 'taboo' subjects for all students of science of his generation in the 1960s, namely *consciousness, qualities* and *animism.* All these 'key' aspects of our world are now 'on the agenda', as he says. The amazing thing being that such subjects could ever have been '*off* the agenda'. Brian Goodwin (who sadly died in July 2009) was an advocate of what has been described as 'holistic science', in which emotion and intuition rank equally with rational analysis of natural phenomena. This was an approach that was inspired by Goethe yet it is as ancient as wisdom itself.

In a discussion about science it is important to ask the question: what is an exact science, after all? Since the advent of what is currently called quantum physics can we confidently measure with accuracy?

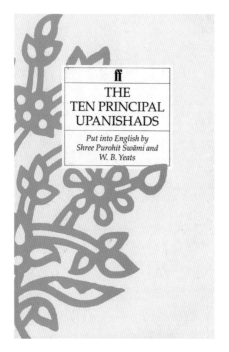

The Ten Principal Upanishads by W. B. Yeats and Shree Purohit Swāmi. One of the most successful collaborations in translating sacred Vedic (Hindu) scriptures.

Maybe the solution lies in our approach to exactitude. This discussion or debate between the *absolute*, the *relative* and the *conjunctive* (which Plato called the soul) is not new. Plato inherited his wisdom tradition via the Pythagoreans who are especially relevant to the problems of our own times.

We have the concept of the *absolute*, without which we have no standard to measure anything that we call relative. As part of this concept Plato, or rather Socrates his chosen spokesman, called the 'intelligible' that which can be understood by the mind or intellect. Then there is the domain of bodily experience which Socrates called the 'sensible' realm and which is now the sole basis of modern empirical science. Neglecting the sensible means drifting in the clouds. Neglecting the intelligible means only indulging the senses.

Our world is the *relative* world, which is constantly changing or 'becoming', as Socrates or Timaeus describes it. There can only be an approximation to any conclusions that arise out of measurements by mechanical instruments that take sensory impressions as their standard. The changing can only measure the changing and cannot itself arrive at certainty.

Thus Plato introduces in his works the mediating importance of 'soul' or that part of our make-up that not only bears life but is the bridge between the sensible and the intelligible. Soul is a clear third factor that is a bridge between the domain of *absolutes* and *relatives*, such as the cardinal numbers on the one hand, and the domain of sensible experience of the numbered on the other hand. The soul is considered the balancing force between two dynamic influences on the human consciousness; this is due to the 'intelligible' finding 'certainties' (or *absolutes*) by which to understand the 'uncertainties' (or *relatives*). There can be no such fact as *un*certainty unless there is *certainty* to measure it by. Thus the relation between the certain and uncertain is the domain of the 'Soul'. And the Soul is a proportional harmony.

The opening discussion of Plato's dialogue *The Timaeus* puts forward two vitally important concepts. First, that of extending ourselves to the utmost to find truth and secondly, the acknowledgement of likelihoods rather than finalities in our seeking of the truth. Timaeus covers his proposals with this important word ('likelihood') as it relates to the subtle nature of how 'like' (or similar) a concept is to that to which it refers. The closer the likelihood or likeness, the nearer we approach to certainty — all the while knowing that our bodily experience can only 'approach' certainty.[8] (There is an interesting relationship here with the Vedic word 'Upanishad', part of the Hindu scriptures. 'Upanishad' means 'near approach', the implication being that we can be helped in *approaching* the absolute truth — not more.) The soul is given the unique role of bringing about a marriage between likeness or likelihood and certainty. This is expressed as the key to the meaning of harmony, actuality being poised between certainty and likelihoods.

In this book we approach the magical domain of flowers, unquestionably a sensorial experience, with the intellectual instrument of one of the 'tools' of certitude: geometry. Or if you prefer, stereometry, the unchanging order of space. ('Earthly' in straight-line geometry, 'Heavenly' in curvilinear or spherical geometry.)

We began this section by questioning the nature of science and scientific method. It is undoubtedly a deep and broad field in which to speculate. The word *science* is simply understood as 'knowledge' in the *Oxford English Dictionary* but qualified as 'esp. of a technical kind'. This does not answer its relationship to mind or the deeper issue of its relationship to consciousness; but it is a reminder that we have inherited via the Latin translation of the Greek *epistemonikos* what we now call epistemology — the theory of knowledge. This volume will be looking at 'experience' and 'likelihoods' more than attempting a theory of how we 'know' or what knowledge is. The human mind takes apart with its analytic habits of reasoning but the human heart puts things together because it loves them, as the traditional saying goes. Evaluation being superior to judgment.

The Pythagorean ethos of the necessity of a simultaneous pursuit of the True, the Beautiful and the Good was profoundly and integrally transmitted in the words of Plato and becomes a reminder, when dwelt upon, that there are *at least* three fundamental aspects of our nature that require nourishment. Thus there is an insistence upon simultaneous concern for all three values to ensure a balanced diet. Our minds require Truth, our values require Beauty, and our wholeness, or soulness, requires the ultimate Good. We will proceed on the basis that whatever else the world of flowers has to offer us is grounded in these simultaneous three values: that they are True, they are Good, and they are Beautiful. Three values that are ever true and life-supporting.

Plato
Timaeus *and* Critias
A new translation by Robin Waterfield

OXFORD WORLD'S CLASSICS

The *Timaeus*, the most influential book, after scripture, for the monotheistic faiths. The earliest scientific cosmogony.

Here the tiny Herb Robert flower blazons its fivefold message, standing
out so strongly against the background. This image also serves as a
reminder of the natural sacrifice which the plant world makes
to the animal world. Life feeding life.

Viewing a Flower Geometrically

Nature, after a certain manner, expresses invisible reasons (or productive powers) through visible forms.

Iamblichus

Geometry is the art of the 'ever-true'.

Socrates

Therefore geometry working with the aid of imagination is able to bring about recollection of the eternal ideas in the soul.

Algis Uždavinys (summarizing Proclus on Euclid)

Having established that flowers can be viewed in many ways and on several different levels, because geometry is the art of 'the ever true' according to Socrates, we have chosen to explore this relatively neglected area: the geometry of flowers. This has significance across all the four levels previously discussed: physical, social, cultural and inspirational.

Our approach cannot and does not attempt to evade the aspect of symmetry — rather it embraces it and draws conclusions from the particular way symmetry is displayed by flower petals. And this in turn relates to the archetypal geometric figure. For once we can take the message from the flower, which speaks to us primarily in spatial terms, in terms of regularity, of figuration and number value, and within the remit of two of humankind's consistently universal languages. These are 'pure number' as arithmetic, and what is classically called 'number in space' as geometry. These two universal methods of communicating are both symbolic (an important but often misused word) as well as technical. (The word *intentional* becomes the philosophically operative word in this case. We are reminded of the work of Sir Jagadis Chunder Bose, 1858–1937, who demonstrated that plants react to human thought, proving that communication does take place between ourselves and the vegetal world.)

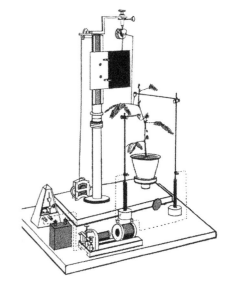

It was not until J. C. Bose developed his recording instruments that the scientific fraternity began to accept his findings. These were momentous. Sadly their implications have been largely overlooked by contemporary botanists.

Pebbles as counters. The most reliable and ancient source of accurate calculation. In the rounded pebbles we simultaneously experience arithmetic and geometry. Plato, Aristotle and Euclid all used pebbles as they had no separate number symbols.

It would be a lost opportunity if we did not consider both the symbolism or quality of number, as expressed by different cultures, as well as the technical meaning of proportional relationships between numbers. However, one of the great dangers of attempting to do justice to both is that we so easily fall into the trap of contemporary Western materialist civilization: an emphasis on specialization inevitably loses the whole. A number is as freely symbolic — even such as the date of birth of each of us — as it is precise in the laws that govern the relationships between numbers. Arithmetic has been and remains humanity's most reliable tool.

Here it is both appropriate and valuable to turn to the rare but potent fragments of the Pythagorean doctrines on number.[9] The reason quite simply is that the Pythagoreans held number in the very highest regard. In fact they held that it was the nearest the human mind could get to participating with the Divine Mind. Be that as it may, the impulse that grew out of the Pythagorean community eventually gave birth to the Platonic and Socratic doctrines, those that saw mathematics, psychology and cosmology as completely integral — each being a reflection of the other: what today we might call 'aspects of universal consciousness'.[10]

We learn that the Pythagoreans probably were the originators of what can be called 'philosophical' numbering. This has also been called archetypal number and it usually refers to the first nine numbers. These were described as the 'incomposite' numbers as each represented a complete quality. It can be useful to think of these numbers as *ness*-es, from *oneness*, to *twoness*, to *threeness*, and so on. This is how Plato can be translated on number *qualities*.

We then follow the Socratic lead in not only taking and thinking of, say, the number three as three units but as a whole, as *threeness*. This is most easily and recognizably done by seeing a triangle as the connection of three points in space at their most balanced, as an equilateral triangle. What better description than a spatial 'threeness'? Flowers with sets of three petals provide examples, such as the Snowdrop (*Galanthus*) or flowers of the Lily family (*Lilium*) — which reveal fundamentally two threenesses. This is the 'quality' of number in space.

When petals are radiant and similar, symmetry is immediately present. Yet each flower that employs this symmetry is doing so in its own particular way. The list of threesomes in petal structures is fascinating. We would like to suggest that in order to view flowers geometrically we take the whole phenomenon of flowering from bud to seed, observing both the two-dimensional and three-dimensional aspects during this time-frame of the flower's life.

These small Pine trees are the natural habitat on Samos, Pythagoras's native island. Pythagoras could not have missed such beautiful examples of spiral structure on the Pine cones of his childhood.

Aα	Bβ	Γγ	Δδ	Eε	Zζ	Hη	Θθ
1	2	3	4	5	7	8	9
Iι	Kκ	Λλ	Mμ	Nν	Ξξ	Oo	Ππ
10	20	30	40	50	60	70	80
Pρ	Σσ,ς	Tτ	Υυ	Φφ	Xχ	Ψψ	Ωω
100	200	300	400	500	600	700	800

The simple letters of the Greek alphabet also had to represent numbers for Pythagoras, Plato, Aristotle, Euclid and others. They were used to give number values to words. This is a deep and technical subject.

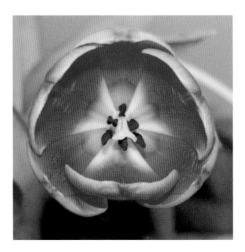

Top. Tulip. The double threenesses are clear yet the centre brings both together in a hexagonal set.

Top right. This Tulip is less obviously composed of two threenesses. Here we see a beautiful triangle in the centre and three yellow triangles which describe a hexagon around it.

From the point (seed), the line (as shoots from the seed) grows both upward and downward. From the light-seeking upper stem arise the planar leaves, and finally beyond these the flowering form delineates a 'solid' or three-dimensional field of action. The three-dimensional aspect will take a profile that demonstrates the movement from bud through to whole flowers. Once this process has reached a certain recognizable stage, we look at the 'face' of the flower in a two-dimensional manner to see how the number shapes and patterns of the petals are displayed. As previously said, the first thing that will strike the viewer's eye will be the symmetry (to a greater or lesser degree) of the petals to one another and to the centre of the flower head itself. This can best be illustrated rather than verbally described.

The Iris became dedicated to the Virgin Mary as the Christian message spread. It is a magnificent celebration of threeness. A flower that deserves the most careful observation.

When fruit is gathered together in this way the hexagonal array is quite clear.

To set up a procedure that will allow us to view these patterns and their geometry systematically, the natural pattern of familiar cardinal numbers will be used. What, however, will not be familiar are the different layers of symbolic meaning for the human psyche. The Pythagorean fragments left by Philolaus are the most reliable as he was believed to have had direct contact with the master — certainly he was at the Pythagorean school in Italy. Firstly it needs to be established that traditionally there is a detectable hierarchy of the nature of number. The three levels in this hierarchy of numbers can be described as:

a) *Archetypal number* — 'incomposite'; complete at nine; limited; 'real' yet summed at ten as the return to unity.
b) *Mathematician's number* — abstract, indefinitely multiple; unlimited; mentally manipulated by the human mind.
c) *The Numbered* — that is, the material embodiment of number 'one of something', four of something, and so on; 'actual' numbering (eventually limited by the third law of thermodynamics?).

The first two aspects of number taken in this way are evidently 'abstract' in the sense that they can be treated without regard to any necessary attachment to actual or physical things. The difference between them is, however, profound. The archetypal numbers remain the qualitative key to numbers *per se.* The mathematician's numbers can (and are) manipulated endlessly without any recourse to qualitative or symbolic value. This last is a technical and quantitative discipline, yet not intrinsically removable from the philosophical perspective of the original archetypal numbers. The numbered or numerical aspect of all materiality can be seen clearly in the atomic structuring of all the physical world as well as a group of, say, seven apples in a bowl.

Having briefly outlined this three-layered hierarchy to number and numbers, it is also important to note that in some cultures we learn that the first two of the nine archetypal numbers are to be regarded differently — as a matter of principle. *One* for instance is to be considered the totality of all number as well as the principle of wholeness and unity — although it equally governs a singular expression of itself (if we can put it in such a way). *Two* also was considered to be primarily the principle of multiplicity and divisiveness, yet also expresses the twoness of all complementarities.

POINT
(seed)

LINE
(shoot)

PLANE
(leaf & petal)

SOLID
(flower & fruit)

The unfolding of the dimensions of space from point to line to plane to solid. The tetrahedron is the first solid.

So in this context the first two 'numbers' are more to be thought of as *principles* of number. This is not far removed from the fundamental triad in *The Timaeus* of Sameness (or oneness) and Difference (twoness or multiplicity) both embraced by their interrelationship or 'Essence' (or wholeness).

What we are pointing to here is that the number we take from flowers can only be a reflection of a universal truth. In this sense they become 'reminders' of the timeless verities. We recommend that the reader turns to the first flower he or she comes into contact with and counts the petals, so lending him or herself to this form of communication.

The five sepals of the Rose disclose the essential symmetry of this dearly loved flower, whose perfume is second to none. This flower has probably the longest history of human cultivation.

TWONESS   THREENESS   FOURNESS

FIVENESS   SIXNESS   SEVENNESS

EIGHTNESS   NINENESS   TENNESS
— which can host a triad within.

Counting numbers and seeing figures simultaneously. The numerical sequence displayed by these
pebbles passes through the cardinal sequence from unity to tenness. Arithmetic is geometry in pebbles.
In the theology of Plato it is the 'First Beauty' that generates all symmetry.

When seen from behind, the sepals and petals of this modest but beautiful
Borage flower dramatically celebrate fiveness in line and plane.

The Quality of Number in Flowers

Because of the gift of sight we have been able to see the stars of the universe and by that we have discovered number.

Plato

Arithmology and geometry are related to discursive thinking and imagination.

Iamblichus

So to return to the flowers: these delicate, mysterious, vulnerable, beautiful life forms (even the most modest of them) can be used as a metaphor for our overall need to satiate our wonderment. We have already suggested how from the point-seed the line-shoots arise, then from the upper stem the planar leaves arise, and finally the flowering form delineates a 'solid' or three-dimensional field of action. This fundamental geometrical pattern follows the ancient Pythagorean simplicity of a seed (a point or oneness) moving to become a shoot (a line or twoness), thus in turn becoming the threeness of the plane (leaf) and finally the fourness (tetrahedrality) of the solidity of flower space and fruit. What is clear is that we have four factors here, yet we have only three dimensions — as a point is either 'non-dimensional' *or* totally dimensional. (See Ananda Coomaraswamy's *Time and Eternity*.)[11] The issue of the oneness of a point we have dealt with elsewhere.[12] The nature of one and oneness is both deep and profound. This concept can be taken as complete simplicity, as well as the greatest philosophical mystery. As with love, it embraces and coheres all.

Oneness continues to be a perennial mystery. We do not think of single-petalled flowers yet the great Goethe proposed, after much observation, that the petals of a flower could be considered as *transformations* of the leaves of the same plant. For differentialists, those disposed to the Aristotelian insistence of categorizing and subcategorizing all of our experience of the natural world, this transformation of petal to leaf is not a 'logical' concept. Yet, at another level, Goethe's perception takes 'at face value' what the flower is actually doing — that is, using its powers of creating two-dimensional features to perform a 'new' and different

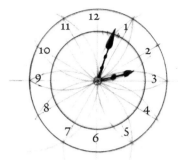

Passing time which is calendar time.
Never again.

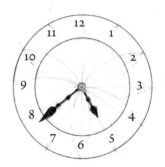

Cyclic time which recurs every 12 hours.
Ever repeats.

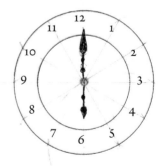

The face's centre is eternity.
It does not move so is non-time.

Time is the mercy of Eternity . . .
William Blake

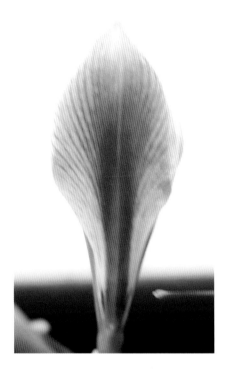

The archetype of all flowering plants. The petal transformed from the light-gathering leaf.

By pairing right and left sides with themselves symmetrically, we can glimpse the inner archetype.

function from the photosynthesis of the leaf, and by doing so creates its own reproductive system. The leaves literally 'flower'!

So the action of one flower reaching up to the light has a beautiful range of functions and what we may validly call 'purposes'. The first can be considered to be the perpetuating of its life. This requires intrinsic life-promoting qualities. These are the need to be rooted in soil; the cyclic and relatively consistent need for water; the movement of air; and the display of features that offer pollen to the insect world, which in turn will seed and regenerate its own species. Above all, reaching out for the life-giving light. And more symbolic purposes could be based on the reactions of the insects, animals and humans who made contact with it, thereby assisting in its reproduction in many cases but most importantly supporting an integral conscious awareness.

We will focus now on the flat or planar aspect of leaves and petals to flowers and follow through the logic of their unfolding. This means both a single flower and each of the petal components of its 'face' or 'head'. The leaves spread out (phyllotaxis) as the stalk rises. We can call this the light-reflecting and gathering area of the plant's 'field' of life. Then we reach the stage in which a completely new function is called for (along with the gathering of sunlight by photosynthesis) — the distributing of the pollen and consequently the seeds of generation. The flower attracts insects to gather the gift of nectar while at the same time picking up the inseminating pollen grain that will go to the next flower. The gift of perfume will also attract certain valuable new relationships, and this is another intrinsically mysterious factor. (What benefit is it to the Rose to give such intoxicating perfume to the human senses?) So if we designate the leaf as a 'oneness' it can be compared to a single petal in the flower. There are flowers which display this quality of oneness of a single petal such as the aspidistra.

One or two flowers will be selected which will offer themselves more easily to geometric analysis. An important feature of leaves is their energy-distributing and structurally-strengthening 'veins'. Key geometric angles in the structure of certain leaves and their veins offer quite challenging issues. These patterns are surprisingly uninvestigated yet clearly evident.

Although we have chosen to describe a leaf as oneness in the sense of its singularity as a flat plane, in fact all leaves have a minimal twoness in the sense of a central stem and a left and right hand side. The mystery of the differences and near sameness of these 'sides' of a leaf is far from easily explained. From the 'oneness', then, a 'twoness' in terms of 'handedness' naturally arises. This can be clearly seen in the case of two pairs of

This is the unique one-petalled flower called Arum. Its spiral unfolding is clearly displayed on the left side of this picture.

This Tulip has two petal types — yet displayed in threenesses. The triangle clearly dominates its form.

Leaves are rarely balanced in their symmetry either side of the centre vein. Here we have paired four sets mirroring one side in each case.

The difference between the two sides brought together as mirror images.

A balanced pair.

Right and left side of the same leaf balanced.

Pure duality.

The Lily, sacred to the Christian faith, is made up of two threenesses or trinities (like the Tulip).

petals. In addition, pairs of leaves across a stalk are quite common even if their sequence above and below is in a more subtle spiral form.

The next in our numerical sequence will be a threeness, so well displayed in the Shamrock or Cloverleaf. This beautiful little plant has leaves which are eye-catching to the geometrically sensitive. The expression of equilateral triangles is beautifully clear.

Three is a powerful symmetry that underlies, as mentioned earlier, the whole of the Lily family. Geometrically, three has to be especially significant, as three points joined become a triangle, *the first shape.* This term comes from *trigon* as used by Plato in *The Timaeus. Tri* is a threeness, but *gon* is more than just the Greek for corner. It also carries the meaning of generation, surviving in the modern English word 'gonads' (male seed storage in the human being) as well as in the word '*gen*eration' itself. The implication is that when a single point relates to two other points in space it becomes a generating angle or a vector (unless they are in a straight line). It is here, in the origins and etymologies of words, that deeper dimensions of meaning emerge. Language has never been considered to be one-dimensional or at least when it is, it is the least interesting or useful or poetic. Symbolism is inherent in each letter of each word let alone the words themselves. Here lies the difference between literalism and poetry. Between solitary meaning and depth of meaning. Herein also lies the potential of reductionism where all is dumbed down into a trivial banality.

Flower petal symmetries advance through fourness to fiveness, sixness and so on — which we will return to later. William Blake spoke of four contributions to wholeness, which he called in his poetic language the four Zoas. Socrates' *divided line* also indicates four levels, which are sometimes called the human Gnostic powers (or powers of knowing). The *divided line* has three divisions and thereby four positions on it. It is divided between the lower two which indicate our response to the sensory world, and the upper two which are representative of the intelligible world (that which our minds comprehend). The two lower divisions are *Eikasia* (estimation), and *Pistis* (belief or opinion) and the two upper areas of consciousness are *Dianoia* (understanding) and *Nous* (knowing or intellection). Rumi, the greatest of Muslim poets, said that there are *seven* levels of meaning in The Holy Koran, 'but only four are available to mankind'. The poet will always prompt deeper meaning.

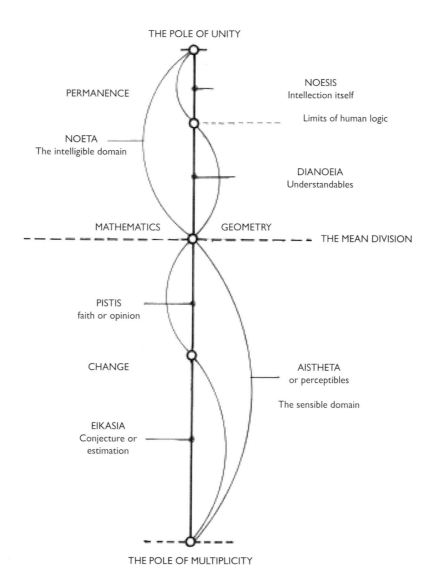

THE SOCRATIC DIVIDED LINE

THE WHOLE

THE POLE OF UNITY

PERMANENCE

NOESIS
Intellection itself

Limits of human logic

NOETA
The intelligible domain

DIANOEIA
Understandables

MATHEMATICS   GEOMETRY   THE MEAN DIVISION

PISTIS
faith or opinion

CHANGE

AISTHETA
or perceptibles

The sensible domain

EIKASIA
Conjecture or
estimation

THE POLE OF MULTIPLICITY

THE PARTS

This is the Socratic vertical line of consciousness or 'knowability'. This follows the Socratic hierarchy of consciousness, which is *eikasia* or conjecture; *pistis* or belief; *dianoeia* or thinking objectively; and finally *noesis* or reaching permanent truths. This statement by Socrates is to be found in Plato's *Republic*, chapter XXIV (VI 509D–511E)

An amazing geodesic geometry displayed by this grain of pollen at microscopic
level. A wondrous world beyond the naked human eye.

Where Does This Lead Us?

At this point it is important to note that this book does not set out to be an encyclopaedia of flowers or even flower types. (There are many admirable versions of these available). What this book sets out to look at is the geometrical order that can be found in flowers. This is spread across the whole domain from the smallest to the largest, from little plants to the giant tree flowers, from the Herb Robert to the Sunflower. It is the *order expressed in the flowering stage* of the plant world that we are drawing attention to — in particular, the effect that this order has, and has had, on the human mind.

Flowering plants not only play an all-important role in creating the air we breathe, without which we could not inhabit dry land. They also have a vital role in the miraculous awakening of human consciousness, one that raises us to the level of responsibility that we define as 'human'. Breath (*pneuma*) is translated as spirit in the authorized King James version of the Christian New Testament. As you, the reader, follow these lines, remember to be aware of the breath you take and maybe recall the debt you owe to the green world. (Plato called the awakening of mathematical consciousness *anamnesis* or remembering.)

As the ancient Japanese maxim goes: 'Civilization was born when the first person offered a loved one the first flower'. Most of us can recall an earliest memory of being struck by flowers or a flower. Flowers are the natural childhood wonder.

Although flowers are intrinsically mysterious, they do reveal 'pattern' to us. The very word 'pattern' is the origin of *recognition. Without pattern we cannot make sense of anything; so pattern is the great reminder. The word itself has associations with French *patron* and Latin *pater*. Pattern is thereby synonymous with recognition and the power of consciousness. This ensures that we are inseparable from flowers as are all subjects from objects. All dualities are mutually sustaining.

The beauty and balance of the face of a flower are the most natural and repetitive reminders of the laws of mathematical and spatial order, the 'laws' underlying beauty. As Socrates says in the *Republic*, 'Geometry is the art of the ever true'.[13] The constant reminder that flowers are emissaries of the 'ever-true' makes what we live in a cosmos not a chaos.

These gigantic, galactic forms affirm the complete universality of the spiral.

The spirals of growth of these cactus spikelets approximate to the golden number sequence.

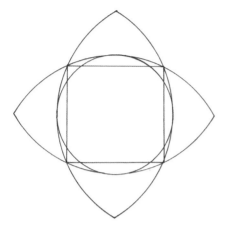

This geometric pattern (squaring the circle) is called the 'Jerusalem' by John Michell. Four arcs, one square and one circle suggest the fourness in one.

When Galileo Galilei wrote in 1623, 'The grand book of the Universe . . . was written in the language of mathematics and its characters are triangles, circles and other geometrical figures, without which it is impossible to understand a single word of it', is this not also true of the inner beauty of flowers? It is one thing to enjoy flowers — another to understand their full dimensions of meaning for us as developing human beings.

Each part of the natural world is inseparably related to each other part — otherwise the word 'universe' could have no meaning. The author is convinced that the rediscovery of complete interrelatedness — or wholeness — is essential at this point in human history. The fractioning of our minds, and the increasing categorization that encourages this, lead obviously and inevitably to the fractioning and divisions in our world view. This eventually becomes fractures in the balance of nature as a whole, giving rise to inevitable conflict, as we are all only too aware today. Surely we have differences, but we are also embraced by samenesses. It is recognizing the conceptual balance between them which can be called the harmony of being.

Herb Robert. This minute flower only goes to prove how powerful fivefold symmetry is on the eye.

A dramatic migrant from the Americas, the Sunflower faithfully follows the sun from which it gets its name.

According to Japanese legend, civilization was born the first time
a loved one was given a flower.

The Thyme Walk at Highgrove House garden,
designed by HRH The Prince of Wales.

TWO

VIEWING FLOWERS FROM DIFFERENT PERSPECTIVES

*It could be said that all trees and plants that are regarded
as sacred owe their privileged situation to the fact
that they incarnate the archetype.*

Mircea Eliade

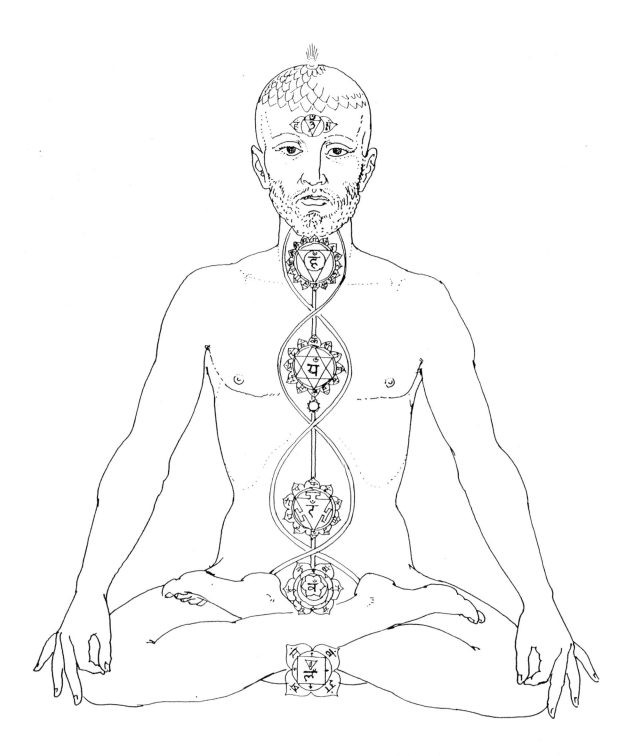

The renowned Vedic centres of the human body called the *Lotuses* or *chakras* relate to our gross anatomy as recognized by contemporary Western science but in themselves are not considered as of a gross nature. This system offers a series of flowers (Lotuses) each petal bearing a letter of the Sanskrit (Devangari) alphabet. Probably the oldest system of human anatomy, it is still very much taught and used today.

The Study of Flowers over Time

That which is being the only answer
The question is its measure. Ask the flower
And the answer unfolds in eloquent petals about the centre;
Ask fire, and the rose bursts into flame and terror.

Kathleen Raine, from Question and Answer

In an effort to be precise about what the natural world actually is, a tendency has arisen to isolate things into increasingly separate and thereby smaller and smaller parts. This grew directly from Aristotle's predisposition to place things into more and more categories —analysis inevitably is the art of differences. 'Whole' became 'things made up of parts' as Aristotle chose to neglect Plato's idea of the constitutional triad of the universe. Timaeus, Plato's spokesman in the dialogue of that name, made it quite clear right from the outset that any discussion or speculation about the origin of our universe should be founded upon (a) *Sameness*, (b) *Difference* (or *Otherness*), and (c) *Essence* (or *Being*). Constant awareness of this triad, simple as it may be, is now considered the only way to maintain an intelligent and harmonious balance between intuition and reasoning. However, this harmonic idea of balance was systematically ignored by Aristotle until it became lost altogether, in the over-zealous quest to become 'scientifically' precise and satisfy human insatiability for novelty and difference. We are, even metaphorically, in an atomic age, having fractured all so-called materiality and thereby experience — and as a consequence have 'fractured' the balance of the human mind as well as our powers of evaluation.

It was Aristotle's favourite student Theophrastus (born c. 372 BC) who put together, no doubt under the guidance and inspiration of his master, the first careful studies of the vegetal world: 'On The History of Plants' (*Historia Plantarum*) and 'On the Causes of Plants'. These were among the first major botanical studies in Europe and formed part of the encyclopaedia project, which was intended to contain all human knowledge. Theophrastus's work included one of the first botanical classification systems, which involved grouping plants into four main categories: herbs, undershrubs, shrubs and trees. It is important to note that the work of Theophrastus required divorcing plants from their normal surroundings

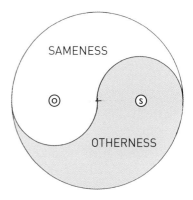

A symbolic version of the fundamental components of Timaeus's cosmogenesis in Plato's dialogue of this name.

The containing quality of essence is the circle itself. This can equally be called 'Being'.

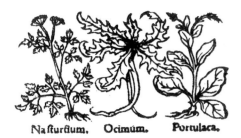

Naſturtium. Ocimum. Portulaca.

Hildegard von Bingen (1098-1179) used a three-level evaluation of the properties of the plant world for healing purposes: (a) the gross or bodily; (b) the psychological; (c) the spiritual. From this she developed her herbal remedies.

Sir Jagadis Chunder Bose showed heroic dedication to his 'science' and an exceptional sensibility. How many 'modern' botanists have carried on his work?

of the soil, climate, insects and other animals, taking them completely out of their natural context. It also required separating and scrutinizing the individual components of the plant, inevitably killing it in the process. In this way, Theophrastus was encouraged to focus on the principle of *difference* which is individuality, and so not only lost the balance and harmony of triadic wholeness *but the life of the plant too*. It is becoming increasingly clear that we have rediscovered the central need to view life as whole. This is reflected in the study of ecology, which emphasizes the *total interdependence* of all things, living or otherwise. Life itself *is* the survival of the fittest as life itself has persisted since its first occurrence in this planet.

Outside Europe, the careful study of plants and flowers can be traced back into the ancient Vedic traditions of the Hindus, which are believed to be over five thousand years old. Dr Sharda Chaturvedi reminds us that the word 'Veda' in the original Sanskrit can be validly translated as 'the storehouse of all knowledge'.[14] In the Vedic tradition, flowers were treated as part of the continuum from the physical to the metaphysical. For example, the subtle *chakras* of the body are visualized as flowers (often called Lotuses) and indicate human spiritual development in this tradition. Careful recording of herbal medicines can be found in these Vedic scriptures. Another of the early classifications of plants can be found in the *Rigveda* (an assembly of Vedic Sanskrit hymns) as well as in the *Atharveda* (another ancient Hindu scripture). The *Rigveda* contains a classification of plants which involves grouping them into *Vrska* (tree), *Osadhi* (herbs useful to humans) and *Virudha* (creepers). The *Atharveda* divides plants into eight classes depending on their characteristics. The fact that these were contained in sacred 'revealed' texts means that their triadic wholeness was maintained. Interdependency is fundamental to the spiritual perspective.

The Chinese also have a long history of plant studies; in particular the development of herbal remedies. The listing of plant and herbal preparations for medicinal purposes dates back at least to the period of the Warring States (481–221 BC).

In medieval Europe it was also considered a valuable idea to produce illustrated 'Herbals', books to help people identify both medicinal plants and those that are poisonous. This preoccupation in the Middle Ages did a great deal to advance herbal medicine. At this time Hildegard von Bingen was an adept in the herbal arts. In her healing work she also took into account the metaphysical nature of plants. This meant recognizing that the higher, more subtle nature of plants is superior to the merely gross and physical, while recognizing that the plant is made up equally of both. She believed that herbs achieved their healing effect by working

through body, mind and spirit because the herbs themselves were also made up of body, mind and spirit. It is clear that ancient China, India and Middle Christendom took the value of flower life to support and 'save' human life as their first principle.

After the first millennium in Europe, owing to the influence of Muslim and Judaic science and the introduction to Aristotle, Plato, Euclid and others, Christendom took a steadily more isolationist approach to the study of plants. This involved studying the physical bodily differences between plants rather than gaining insight into the value of plants as a whole; particularly the benefits that plants have on human health. This approach developed to the point that it was deemed necessary to uproot (and so kill) each specimen in order to study what is usually unseen, the inside and the roots. The respect for *Essence,* which is clearly analogous to Life, was fundamental to the traditional perspective and this was steadily being pushed aside by the schoolmen in favour of human speculation (literalism) and verbal argumentation (logic). The farmers, gardeners and husbandry men, however, were not affected by this emphasis on exercising the human mind — they knew what was good for humans and animals as part of an uninterrupted stream of earthly wisdom, inherited from time immemorial. Agri*culture* was an integrated mutual understanding between people and the natural world. The wisdom was transmitted by doing rather than arguing about; an artistic sensing rather than a rationalization.

Here we see humanity at its wisest in relation to the natural world. This is 'agriculture' as 'agriwisdom', the very opposite of 'agribusiness'. The wisdom in these carefully planned crop fields demonstrates a sustainability that has worked for thousands of years. We can live in respect and harmony with nature or we can display uncurbed greed and proceed to destroy the natural balance.

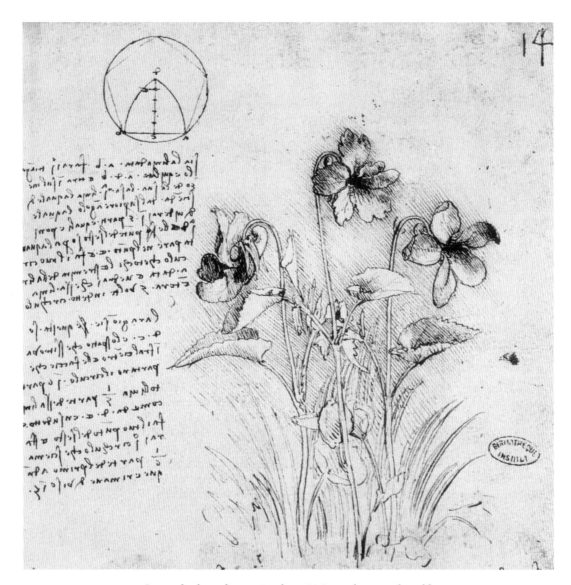

Leonardo showed exceptional sensitivity to the natural world
— and recorded it. The pentagonal geometry of the fivefold
Violet is evident above.

In the period some call the Renaissance (fifteenth century) there was a re-emergence of the classical philosophical doctrines, including the works of Plato, Aristotle, Proclus, Euclid, Plotinus and many others. These had been transmitted into Europe by the Muslims and the Jews through North Africa and encouraged a much closer scrutiny of the combined beauty and truth of the plant world. Hence we see the prime example of the skill of artists such as Leonardo da Vinci whose plant drawings show an immense sensitivity to, for example, the spiral nature of their unfolding, which as a principle can be taken as a good example of *Sameness* in the plant kingdom. Leonardo's drawings were also magnificent examples of the beauty of the differences between plants. He maintained the balance between *Sameness* and *Difference* as he no doubt had learned of Plato's doctrines — which were available to him at this time.

The modern 'botanical' study of plants has its origins in the early part of the sixteenth century in Europe. The word originates from the Greek *'botanizein'* which meant to gather plants. Today, botany is described as *'the scientific study of plants, including their anatomy, morphology, physiology, biochemistry, taxonomy, cytology, evolution, and geographical distribution'*. This set of differences in approaching plants speaks for itself.

There are less well known contemporary writers who have become particularly conscious of the importance of the Greek philosophers. One such is Jeremy Naydler who is very aware that modern empirical science is losing its reference to wholeness. In his seminal pamphlet *Perceptions of the Divine in Nature*, Naydler talks of the 'heart' of the Lily as a metaphor for the way in which we study flowers.[15] The heart of the Lily refers to the archetypal essence of the Lily as well as our inner response to it. In his work he traces the recent scientific line of enquiry in which the pursuit of 'inner meaning' has become disastrously confused with the mechanical analysis of plants into smaller and smaller parts. As a result, considerations of the (normally invisible) gene have become more important than experiencing the flowering plant as a whole. We too easily *dismiss what the flower is communicating* to us, even though the pioneering scientific work of Sir Jagadis Chunder Bose in the 1920s had clearly proven to the British scientific community that the flowering world responded to human consciousness. The multiplying of differences could aptly be called the science of the wrong track, because it is the opposite of pursuing wholeness. It is this very urgent need to grasp the meaning of the whole which gave birth to the contemporary science of ecology. Without a basic reference to wholeness how could we conduct a responsible science? Surely interdependency and collaboration in the living world will lead to a greater understanding and enhance the necessarily intrinsic harmony in the world, benefiting all.

Leonardo's sensitivity to the spirals of the plant world is clear in this study — especially in the leaves of this Star of Bethlehem.

The network of subtle anatomical channels according to the Vedic tradition is
not empirically acknowledged by physical scientific evidence but is fundamental
to the spiritual anatomy of Hinduism — just as the acupuncture meridians are
fundamental to the Chinese system of anatomy.

The Traditional Perspectives or How We Could Regard Flowers

The elders were wise. They knew that man's heart, away from nature, becomes hard; they knew that lack of respect for growing, living things soon led to lack of respect for humans too.

I am going to venture that the man who sat on the ground in his tipi meditating on life and its meaning, accepting the kinship of all creatures, and acknowledging unity with the universe of things, was infusing into his being the true essence of civilization.

Chief Luther Standing Bear from the Lakota Sioux

Curiously, the 'whole' is one of the most difficult concepts the human mind can engage with. Why? Because we either drown in detail or we become aware of the huge areas of our ignorance, yet we instinctively know the whole is fundamentally a great simplicity and ultimately oneness itself. This is as true when viewing flowers as it is for the natural world itself. The doctrine that we are a microcosm is the traditional wisdom, in one form or another, for all traditional civilizations.

There is so much information available today via mechanical and electronic recording and storage systems that it would not be an overstatement to suggest even the most highly specialized 'scientist' cannot keep up with all the information generated within his or her own specialized field, let alone relate it to the whole.

This brings us to the reason why the human family has in its wisdom developed philosophical approaches to human consciousness and knowledge. Philosophy itself as a word comes from the ancient Greek *philo*, 'love of', and *sophy*, 'wisdom': in short the philosopher is a 'lover of wisdom' a term often credited to Pythagoras himself. This was the highest 'calling' or 'career' a person could choose according to Plato, as it was the only one that could give a meaning or purpose to existence. This he unfolds brilliantly in his dialogue the *Republic*. Wisdom is as much right action as it is right thinking and right being. Left to itself, the human mind tends to confuse itself through over-elaboration and one-sided concerns rather than seek the wisdom of intuition.

This essentially nomadic dwelling of the North American Indian peoples was axiomatically in harmony with the natural world.

I wonder if the manipulator of the fate of this Tulip even paused to consider that it might have had feelings (as proved by Sir J. C. Bose) about its final appearance? What is our moral position regarding such blooms?

At birth, we find our senses assailed in all directions with effects that we (necessarily) need to assimilate. This first level of knowing, which we call *sense data*, is the first level of 'knowing' through the senses. It is a confusing multiplicity of sights, sounds, feelings, smells, tastes and so on, pure difference. From sense data we begin to employ what we call *re*cognition to acquire 'information': which most interestingly means quite literally 'to give form to' or 'to give a character to'. This is then the basis of that which can be recognized again: the employment of sameness. After *sense data* and *information*, we can call the third stage *knowledge*, which means the knower or recipient has become conversant with the subject in hand. It refers to an ability to recognize or distinguish and agree shared knowledge in order to establish truth. So we have a ladder from *sense data* to *information* to *knowledge*, each adding a qualitative difference to the previous stage and equally each progressively simplifying sensory experience. From *knowledge* comes the clearly disciplined fourth stage, which is the distillation of knowledge. This process can be described as seeking the essential values which offer a unifying wholeness. This is the basis of what we call *wisdom*.

This qualification or evaluation of knowledge to find the most pertinent aspects (leading towards the timeless) becomes *wisdom*, that is philosophy in its purest sense — wise thought integral with wise action. Wisdom can thereby be valued as the goal of knowing anything fully. Equally, comprehension of the importance of the flower follows this scale.

For each level on this 'ladder of knowing' we can recognize a time factor. Sense data is immediate and constantly changing. Information works on a somewhat longer timescale; fashion comes into this category, as do the news media. Knowledge is of a much longer timescale and usually refers to qualities that can be referred to over many, many years, even millennia. Finally, true wisdom refers to the unchanging qualities that are timeless or as Socrates called them the 'ever true'. Such tools at this level are the 'universal languages' established in the Pythagorean curriculum or educational system of number such as arithmetic, geometry, harmony or music, and astronomy. These 'arts' are true for all at all times, therefore not limited by passing fashion or culture — hence their essential value. They are the most reliable tools for assessing the passing or changing world of experience and essential for finding wholeness, the purpose of life and existence itself.

This scale relates to the four perspectives that were described in the Introduction. The four perspectives represent the traditional way to view any subject, particularly flowers. These perspectives can be called:

a) The material — 'what does it appear to be made of?'
b) The social and psychological — 'how does it affect human life?'
c) The cultural and mythological — 'how has it been embraced by each civilization?'
d) The inspirational — 'what is its ultimate source and goal?'

The material world is studied from sense data, the social sphere through psychological experience and insight. The cultural world embodies the knowledge and conventions of each civilization or culture. The inspirational world offers wisdom and the timeless guidance coming from the 'ever true'. This last is also what comes under the word 'revelation' or the 'inspired scriptures' of each spiritual tradition. Most conflict and misunderstandings between people arise due to a lack of understanding about which perspective they are taking at any given time.

Another wonder of the plant world is the management of seed migration by means of wind. See illustration of whole Dandelion sphere on page 29.

In attending to any subject we can move up the fourfold ladder or *descend* down it. This can be called the twofold motivation. On the one hand, the desire to find the overall truth and seek samenesses in the timeless principles. On the other hand, the *descent* that naturally occurs when one is drawn into constant concerns such as financial security, social honour or an attachment to sensual and bodily needs — in short ego concerns that are by nature differentiating, the actual nature of individuality, our separateness.

In our concern for the plant world we see the same possibilities. On the one hand, seeking the principles within and behind flowers and the timeless principle of flowering itself. On the other hand, seeking the exploitation of usually cut and genetically-engineered, even patented, flowers for financial gain with a disregard for the sensitivities of the flowers themselves. The gardener who lives for beauty (and the flowers themselves) will be unconditionally fulfilled.

The *relationship* between flowers and ourselves is still an astoundingly unresearched and unacknowledged area, even after the revolutionary work of such as Sir Jagadis Chunder Bose in the 1920s as already mentioned. Genetic engineering for financial gain is far the most predominant activity in the contemporary industrialized world. And who is continuing the important findings of Bose?

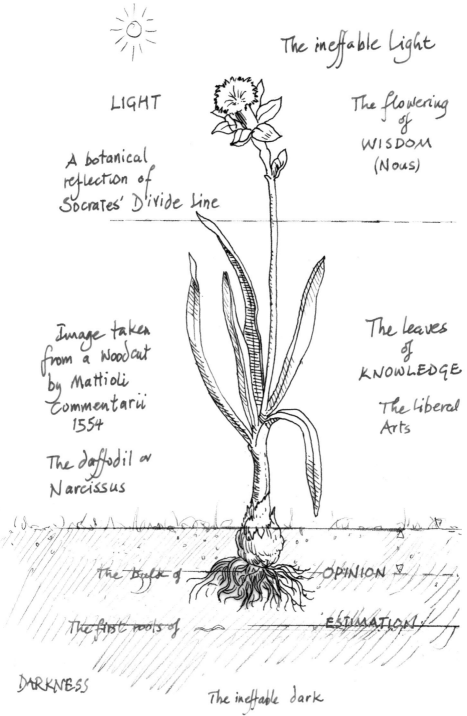

The ineffable Light

LIGHT

The flowering of WISDOM (Nous)

A botanical reflection of Socrates' Divide Line

Image taken from a Woodcut by Mattioli Commentarii 1554

The daffodil or Narcissus

The leaves of KNOWLEDGE

The liberal Arts

The Bulb of OPINION

The first roots of ESTIMATION

DARKNESS

The ineffable dark

Here we use a natural form, the Daffodil to illustrate symbolically the fundamental Socratic doctrine of levels of 'knowing' or consciousness. The flower has four distinguishing functions: roots, root stalk, light-synthesizing leaves and flowering itself. These follow the Socratic hierarchy of consciousness, which is *eikasia* or conjecture; *pistis* or belief; *dianoeia* or thinking objectively; and finally *noesis* or reaching permanent truths.

We *can* choose to see the phenomena of flowers, their life and flowering as an inspiring ingredient of our planetary home. This does not mean we cannot say in huge detail what floral life does and how it behaves, with all the extraordinary details of its inexplicable integrality and cooperative activities. We are only at the edge of understanding the deeper meaning of life, as we have almost completely invested our time in scientific laboratories analysing the minutiae of microscopic details at cellular, genetic and atomic levels. *How life* works is not *what life is*, nor is such an enquiry even concerned with natural ways to enhance it. And what is for certain is that life is a noumenal principle which literally animates matter, matter that we call 'dead' when life abandons it.

Life lifts as materiality drags down. Levity as a word has virtually dropped out of common usage; but we all experience gravity and are 'held down' by it. Levity gives us hope, purpose and a reason to be alive. Gravity, at best, keeps us grounded; at worst sees us to our graves.

Flowers, as both phenomena and noumena, offer their being as forces of levity through their Beauty, Truth and Goodness. It is good for us to regard their existence with gratitude and awe. (Even Goethe often repeated his comment: 'Everything in nature lives by give and take'.)

This beautiful small white flower, the Tradescantia, demonstrates so well its scource in three-fold symmetry, see p.215.

In our view this is one of the most valuable discoveries by the great Swedish botanist Linnaeus.
The flower clock has more to say to us than merely 'telling the time', however valuable it is.
'Flowers have always had important things to teach us about time.'
M. Pollan (from *The Botany of Desire*)

The Material Perspective

There is a delicate empiricism which makes itself utterly identical with the object, thereby becoming true theory. But this enchantment of our mental powers belongs to a highly evolved age.

Goethe

Considering the whole of any plant has a great deal in common with considering any integral form of life. The fact of life is that it takes hold of organic and inorganic material and transforms it into the light-seeking impulse employing rhythms we recognize as living. This has been called the 'return to unity' impulse, based on seeking light. It sustains through continual renewal.

Having outlined the above there are other environmental aspects such as soil condition, geological elevation, wind conditions, rainfall, and possibly as important as these, the method of pollination. Methods include wind and the insect world from bees to butterflies — with a range of flying or crawling creatures in between. All are instrumental in the fertilization of each flowering plant and even the proper oxygenation of the roots. They all confirm the principle of collaboration and clearly the interdependency of all life.

The life-span of a flowering plant follows a rhythm from season to season until it is 'reborn', usually about a year later although there are many variations. Flowers can continue to bloom through a summer — even a whole year in certain climes. This means that the time factor is critical in making any study. What becomes apparent to even the most elementary study is that a plant embodies and responds to rhythms. The most evident of these is the diurnal cycle of twenty-four hours. It is important to remember that this is a cosmic rhythm and function which affects all things on earth. This once again confirms that organisms pulsate rhythmically. Life is rhythmic. Our heartbeat most emphatically confirms this.

The turning of the earth every twenty-four hours affects all living things on its surface. This is signalled by the arrival of light at dawn, or even before dawn, and the setting down of light at dusk. Many flowering plants are solar-following as well as solar-opening. The light from the sky, direct sunlight or other, is the prime stimulant for a plant to flower and grow. Physically light is the ultimate nourisher (and timekeeper).

Many plants will however be equally at home in shade which means

An eclipse of the sun. A singular miracle of life on earth is that we find the sun is exactly the same size as the moon. When we have no rational explanation we call such an experience 'a coincidence'. Both sun and full moon are perfect circles to the human eye.

diffused and lowered light levels — on a forest floor for instance. There are many, many survival and thrival strategies to be found in the plant world. The major rhythms being the seasons of the year and the local climate.

That the sun and the moon are the same size in our sky was taken by the ancient sages as the most precise sign that we live in an ordered cosmos and not in a chaos. When the moon eclipses the sun, this precision of size is dramatically revealed. It also bespeaks deeper things to do with the significance of our perception, as well as the *place* of human consciousness in the whole scheme of things. The rhythm of the year is struck between the sun and the moon. There are virtually thirteen full moons in a year and these are distributed within a twelve month cycle which is signalled by a particular background pattern of fixed stars. This we have traditionally called the zodiacal belt. Its division into twelve different 'constellations' is where the word 'zodiac' comes from. *Zoë* is 'life' or 'living' in the original Greek. The constellations held significant life symbols for each civilization as well as being time markers.

The planetary 'family' which we are within is defined by the fact that

The zodiacal circle. The division into constellations is caused by the co-ordinated rhythm of the sun and the moon. The night sky is divided into twelve mansions. Time and the seasons are measured against these cosmic rhythms and divisions.

Opposite page, right.
These represent time diagrams of the geocentric movements of the five planetary bodies in relation to ourselves on earth. The plotting of these scientifically accurate paths is attributable both to ourselves and to each planet revolving around the sun. The 'floral' aspect of the orbits is striking. The ancient Chinese were the first to record these geocentric patterns.

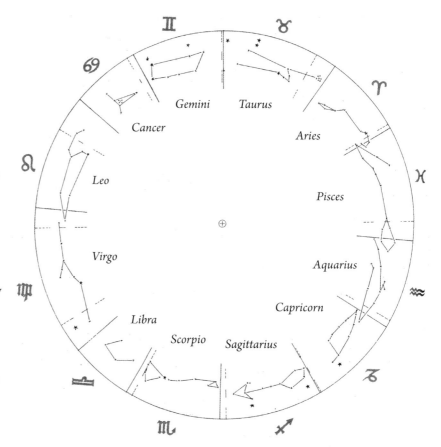

the planets are able to be seen by the naked eye, and each of them creates special rhythm against the night sky. The phenomenon which some call 'retrograde' motions, Plato chose to call the *choreography* of our universe. Each visible planet traces out its own form in time. These exhibit extraordinary symmetries and thereby come into the same *laws of existence*, as do the flowers on earth. It is the similarities that hide clues as to how the planetary rhythms may affect life or lives on earth.

Each planet has its own rotation period, ours on earth being 23.934 hours. Each therefore has its own average day length. Each has an orbital period in relation to our 365.256 day period. We all accept that the moon with its orbital period of 27.3217 days affects not only all the oceans on earth but the liquids within the sap of trees and even plants, as well as animal and human fertility cycles. There is no hesitancy that the light of the sun is the light and life of the world. Day and night and the equinoxes and solstices characterize the fourfoldness of both our twenty-four-hour

The continuous linear diagram of the relationship between the earth and the planet Venus. How could one not see a flower in this time diagram? (See John Martineau's *A Little Book of Coincidence* detailed in the endnotes.)

MERCURY

VENUS

MARS

JUPITER

SATURN

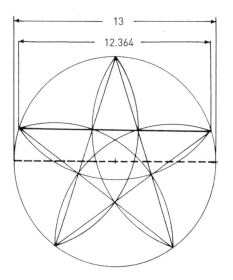

This is another way in which the star pentagon can offer us information about cosmic rhythms. The pentagonal flower within the circle offers us a precise measure between the solar rhythm of the year and the lunar rhythm. (With acknowledgments to Robin Heath.)

cycle of dawn, day, dusk and night and the rhythm of our single year. Not only this but they determine the seasons of our plant-life rhythms. The other planets (or gods as far as the most sophisticated ancient Greek philosophers, including Thales, Socrates, Plato, Aristotle and Pythagoras, were concerned), were studied in ways that we as moderns would probably not recognize. Recent studies in these rhythms by such as the highly developed Astronomical School at the Goetheanum in Dornach, inspired by the philosopher Rudolf Steiner, have revealed a whole new field of study. Each year the Anthroposophical Society, (uniquely in the scientific world as far as this author is aware) publish the *empirically* factual scientific patterns traced in the sky by each of the major planets. The results of seeing the time-traces of these major inhabitants of our universe are not less than astonishing, particularly the flower-like patterns seen from the geocentric (our own) natural and empirical view point. They demonstrate Plato's use of the word 'choreography' perfectly — as well as reminding us that philosophers such as Plato did *not* publish all they knew. The word 'publish' means to make available to the whole public. Pythagoras published nothing, but his *School* influenced the next thousand years of Greek thought. When pressed, Plato said 'I am afraid much must be left unspoken' or 'I am omitting a great deal'.[16]

This is not the place to unfold the possible implications of which and how these rhythmic patterns might affect life on earth, particularly plant life. But we recommend that the serious investigator refer to the long-standing work of the Anthroposophical Society in Dornach as well as the writings of Rudolf Steiner, a man deeply influenced by Goethean science.

We can only, in this instance, display *some* of the symmetries involved, including the flower-like time-traces that each planet makes in relation to us on earth. We could call them 'planetary flowers' in time as experienced or seen from the earth.

Also illustrated are some of the extraordinary coincidences that occur between the mean orbits of our planetary system and some of the mean physical sizes of key planets. They are pure stereometry or the geometry of light.

Pattern, as we have discussed, is the foundation of a cosmos in contradistinction from a chaos. Order, proportion, symmetry and harmony are qualities not things. Things may *participate* in these profound qualities, (may be governed by them) but the qualities themselves remain above the physical. The *Oxford English Dictionary* describes metaphysics as a higher science than that of physics. It is not a case of what came first, the chicken or the egg, as both these are 'things'. It is rather that when matter emerged, principle was already there. Otherwise no order could

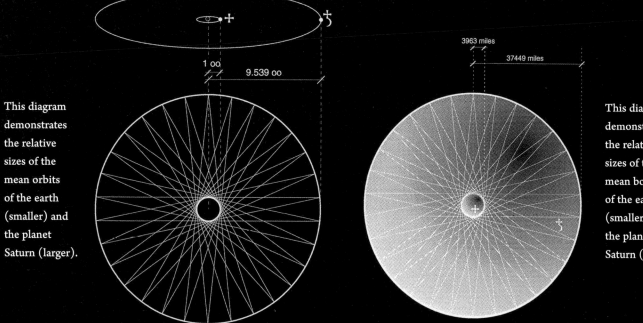

1 oo   9.539 oo

3963 miles   37449 miles

This diagram
demonstrates
the relative
sizes of the
mean orbits
of the earth
(smaller) and
the planet
Saturn (larger).

This diagram
demonstrates
the relative
sizes of the
mean bodies
of the earth
(smaller) and
the planet
Saturn (larger).

These 'coincidences' were uncovered by John Martineau while he was an MA student with the author in the VITA department of the Royal College of Art, London. They demonstrate an unquestionable order to elements of our solar system. We can choose to live within an 'accidental' universe or a universe of profound order.

0.3870 oo   1 oo

1.618   1

These two circles
represent the
mean orbits
of the earth
and the planet
Mercury. Their
relation to
the golden ratio
is remarkable.

This shows the
relationship
between the mean
bodies of the
earth (larger)
and the planet
Mercury, also
in a golden
relationship
as with the
orbits opposite.

This beautiful affirmation of the golden relationship between the bodies and orbits of the Earth and Mercury comes from seeking the archetypes. Here John Martineau has sought the 'mean' values in the measurements — a balance between the extremes or their archetypes. The 'golden mean' is hereby shown to be a product of the 'universal order' in our solar system and certainly not our invention.

have ensued. This is a principle shared by the traditional wisdoms of humankind.

Naturally we can say 'without matter no principle', but also more importantly 'without principle no order in matter'. We can take neither for granted, they are obviously integral. The current author finds the following set of principles quite adequate to cover the conditions of existence and wishes to acknowledge F. Schuon as the source:

Space: Which gives birth in human consciousness as geometry and measure.
Time: Which gives birth in human experience as duration and rhythm.
Number: Which gives birth in the human mind and body as arithmetic and quantities as well as qualities.
Form: Which gives birth in the human perception as shapes, limitation or demarcation.
Substance: The underlying cause of the arrival of sense impressions in humans — eventually becoming materiality.

In the world of flowering plants these conditions are vitally important. If we are to make a holistic comprehensive study, we need to employ consideration of all of them both singly and in collaboration.

The Social and Psychological Perspective

Listen, Life is one, it has no beginning and no end. The source and goal live in your heart.

J. Krishnamurti[17]

For the following information I would like to thank my son Dr Matthew Critchlow who offers a holistic health service called 'Wellness Insights' and 'Thrive' and is best informed in this area.

Is it a reflection on the way modern science has developed that there are virtually no accepted scientific theories to explain the existence or value of positive human emotion? We are all aware that modern psychology began with studies of pathology and negative behaviours and emotions. How many scientifically-accepted theories or even studies exist on positive human emotion? Only since the advent of the 'positive psychology' movement in the late 1990s (inspired by the work of Martin Seligman and colleagues in the US) have modern psychologists invested any significant time or energy into the study of positive emotions.[18] In this instance, positive psychology has three concerns: positive emotions, positive traits and positive institutions (such as consensus-led, strong families and free expression, these being the psychologists' value terms and clearly requiring further explanation). At the heart of this relatively new discipline is a belief that the understanding and promotion of mental wellbeing is far better served by studying people who are happy and relatively contented (experiencing positive emotions) than it is by relying on the study of people who are mentally unwell (experiencing negative emotions such as depression and anxiety). However, serious issues about definitions will inevitably arise.

One recent study into positive emotions was undertaken by a team led by Jeanette Haviland-Jones. Their paper is called 'An Environmental Approach to Positive Emotion: Flowers'[19] The opening statement claims that for five thousand years humans have cultivated flowers *although there is no known reward for this costly behaviour*. (Italics added.) Note that the word 'reward' can only be used in a financial context these days.

Three stages in the social experience of wearing flowers:

1. The wonder of wearing a Rose garland for the first time.
2. The pleasure of being told how nice the ensemble is.
3. The resultant joy of wearing a garland for a special occasion.

This begs many important questions. However, the study proceeds to demonstrate, in empirical scientific experiments, that flowers are 'positive emotion inducers'. This is very much to be applauded, as it obviously supports a universal piece of common sense and common experience.

Three studies were conducted by the team. The first considered the effect that flowers had on women who received them as a presentation. The second considered flowers presented to both men and women in a lift (or elevator). The third considered a flower presented to elderly participants (of 55 years and older). All had positive 'mood reports' and the elderly group had 'improved episodic memory' after receiving flowers. What the team found was that flowers have immediate and long-term effects on emotional reactions, mood, social behaviour and *even memory enhancement* for both males and females. This team then significantly records: *'there is little existing theory in any discipline that explains these findings'*. This once again highlights the lack, in contemporary empirical science, of philosophical knowledge or even a holistic overview. Without a hierarchy of levels the 'empirical' is grounded in its own restrictions which are nominally only experience-based. As the ancient Chinese sage said: 'The logicians only settled terms but lost common sense'.

By adopting the ancient wisdom tradition which uses at least a fourfold 'ladder' we can have a tool to review the shortcomings of modern empiricism. We will summarize the fourness of consciousness by amalgamating three or four different traditions. We have already touched on this fourfold ladder of knowing. So to give us a conscious foothold on this ladder, we suggest that the next rung or level of consideration be called the psychological, which is also the social as well as the symbolic. It is this level that our team of Evolutionary Psychologists have entered upon and also the reason that they find 'little existing theory' to refer to — that is, within their confined area of definition. A good example of a text that they would surely find useful is the *Consolations* of Boethius. This is a most welcome journey back into the time-tried wisdom traditions. The next rung of the ladder will be '*cultural*' or *civilizational*: the considered repository of human experience (which is discussed in the next section). Overseeing this is the perennial journey toward an all-embracing and simplified wisdom, which we have chosen to call the *inspirational* rung of the ladder (described in the final part of this chapter). Beyond this is what has been traditionally and inevitably called the inexpressible or ineffable. This is the experience of a oneness or unity that is quite beyond description, as description itself cannot but be the division between that being described and the description itself. (Plato's 'must be left unspoken'.)

The above research has successfully bridged the all-important step in the ascension of understanding, supporting the obvious fact that beauty in the form of flowers evidently does affect people most positively. In fact, they talk of long-term effects and even memory recall. Could it not be that the positive rewarding effects of flowers suggest they have had a vital role to play in bringing out the finer sensibilities and consciousness that we now define as human? After all the plants were here on earth millions of years before we arrived. The discovery of pollen in the graves of the Neanderthal suggests that Neanderthal humanity treasured flowers, even before humanity became what is now popularly defined as *Homo Sapiens*.

Flowers are the social bridge between the generations.

Marriage is the symbol that puts together and reunites human life with the whole of Nature. Flowers attend all weddings.

The Lotus flower has been held sacred by the greatest number of people over a long history, from Egypt to India to China and Japan. It is an exceptional flower to say the least. Here we see the central 'point-number' nature of the seedpod. The 'seat'of most images of the Buddha, this flower is believed to have once inhabited the whole surface of the earth.

The Cultural and Mythological Perspective

The righteous shall flourish like the palm tree: he shall grow like a cedar in Lebanon.

Psalm 92:12

For my dreams of your image that blossoms a rose in the deeps of my heart.

W. B. Yeats

The Buddha almost invariably sits on the seedpod head of a Lotus flower.

That the Lotus has been called 'enlightenment itself' is not surprising as the Lord Buddha sits on the seedpod in the centre of the unfolding petals quite without disturbing the flower! An example of profound symbolic meaning. As the flower has arisen above the mud and water of the material world so the 'Buddha Nature' rises above it. From light to enlightenment.

As discussed in the previous section, the next rung of the ladder in our understanding of the universe is the 'cultural' or civilizational level, which includes myths and mythology. True civilizations or cultures could be said to work only on a cycle of thousands of years.

Sadly, the word mythology provides a good example of how modern fashions can devalue or completely debase a word. (As William Blake said, if you want to destroy a civilization first debase its terms and words). The proper study of mythology can be described as entering into the metaphysics of a given age or civilization. For the present author mythology represents modes of description that are outside of historical logic. Myths deal with *truths that are timeless* and require cladding in the language of time to bridge the gap. So to many people they are merely stories.

The current meaning that has been attached to words like 'myth' is that of 'untrue' or 'irrelevant'. The loss of transmitted folk wisdom, children's tales and cultural stories of heroes and heroines, provides a powerful example of the loss of our inner knowledge. Myth means rather 'more true than can be explained in words' in the sense of a truth that is received inwardly but not subject to the scrutiny of 'factual' analysis. Certain inner truths are only transmittable through 'likely stories' as Timaeus, Plato's spokesman, says.[20] 'Likelihoods are all that are available to humanity', implies Timaeus, when it comes to investigating the ever-changing phenomenal world. (Quantum physicists would most likely agree today.) Hence to gain access to deeper truths the construction of myths was the method used. In scriptures these are called *parables* or *truths received* on a different level.

G. H. Mees is probably the most informed and inspired contemporary transmitter of the sacred traditions in their deeper symbolic clothing. His three monumental works, *The Book of Battles, The Book of Signs* and *The Book of Stars* are the richest volumes in comparative sacred traditional symbolism available in English.[21] His 'style' will not suit all investigators but his clear integrity can greatly help in approaching his unusual and unfashionable way of transmitting knowledge.

In *The Book of Stars* (page 169) Mees stakes his position with great clarity, leaving readers to take it or leave it as they choose. He states:

> When traditions are no more taken symbolically, when their esoteric life ceases to predominate over and guide their exoteric [literal] form, they degenerate either into monstrosities or into pale ghosts of literalism.

He warns of the current state of degeneration of the inner meanings of the great traditions and the accompanying conflicts that we all are only too aware of. His term for our times is '*this Black-out twilight*'. Having lived through two world wars, Mees had good reason to know 'his times'.

Two other recent authors who have written deeply on the essential necessity of acknowledging symbolism are Dr Martin Lings in his inspiring small volume *Symbol and Archetype* and Mircea Eliade (particularly in his *Studies in Comparative Religion* and *The Sacred and The Profane.*)

It has been claimed that the plays of William Shakespeare contain mythologically and psychologically the whole gamut of possible human relationships, their lessons lying in the historical acting out of all the perennial human predicaments. We have chosen just a few from the grand and intriguing mythologies connected to certain plants and flowers to make our point.

Probably the most world-renowned mythological flower symbol is that of the Lotus —particularly in relation to the Lord Buddha (to which we will return later). The greater number of images of the Lord Buddha present him with his body either sitting, standing, or lying on a Lotus. However the Lotus in each case is not crushed. So what does this impossibility mean if we accept that his physical human body is, like all human bodies, of a considerable weight? The answer quite simply is that the essential Lord Buddha is of a subtle and universal nature, one that has become released from the attachment of materiality and is thereby a greater metaphysical truth, and in being so, weightless. In these instances the imagery has become mythological and representational simultaneously.

The Lotus is also a mythological symbol of the flowering of the human soul. In both Buddhist and Vedic tradition they are called 'flowers', 'wheels' or 'chakras' in Sanskrit and each of these 'lotuses' has a particular symmetry and each is related, according to some traditions, to one of the seven heavenly lights or archetypal planets. Such understanding also emerges in Christian Europe in the work of Johann Georg Gichtel, a student of Jakob Böhme (1575–1624). This is a very large subject and separate studies are recommended to those who are interested in this aspect of flower mythology. This is not to diminish their importance but rather to emphasize their purely mythological reality in this perspective.

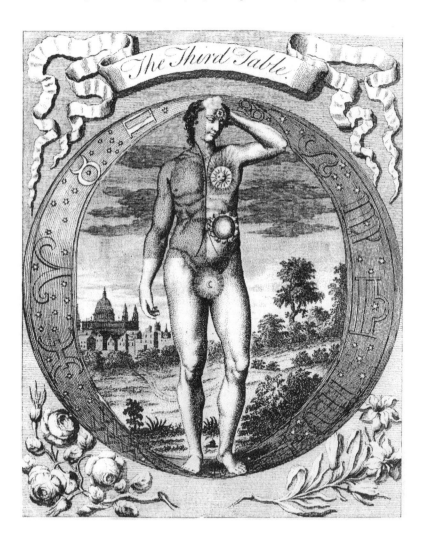

Gichtel, a student of Jakob Böhme, illustrates the tenets of human subtle anatomy according to his master's teaching. Each 'centre' is characterized by one of the planets. The closeness of this image to the oriental versions of the chakras or Lotuses indicates that such subtleties have been available to all spiritual traditions.

The central door post of the south porch of Chartres Cathedral is carved into a statue of Jesus holding the Gospels in his left hand and blessing with his right hand raised. However, the significant aspect of the crown is that the interior is made up of Roses and Rose leaves, while the exterior is once again the Heavenly City or the Heavenly Jerusalem. Below this figure are carved the bakers who financed this column. It is particularly noteworthy that the chief baker is himself wearing a floral garland. Hence we can conclude that the flowers represent blessings in both circumstances.

Wreaths of Plant Life: The Symbol of Attainment

The centre is still and silent in the heart of an eternal dance of circles.
Rabindranath Tagore

The captive flower in the king's wreath smiles bitterly when the meadow flower envies her.
Rabindranath Tagore

The language of floral wreaths or garlands is typically mythological and symbolic. A wreath is usually woven into a circular form which enhances the inherent symbolic 'power' of the plant used. Twigs, leaves and flowers are involved. It must not be missed that this crowning glory was also symbolic of the crown *chakra* flowering in Vedic symbology, a symbol of the highest spiritual achievement while in human bodily form. A wreath is also the 'halo' of plant life.

J. Dierbach wrote of Ivy wreaths that 'ivy was used to crown poets as a sign of their keen intellect and subtle minds; this is why Albertus Magnus said that Alexander the Great crowned his officers with ivy. Alexander the Great himself wore the same to celebrate his victory in India'.

The Hawthorn was venerated by the Celtic peoples, especially in Ireland where a 'lone bush' or 'sentry thorn' was considered holy to the 'little people' or fairies. Farmers laboriously worked around these, particularly if they were also on a 'wrath' or mound. In the 1960s the present author and his family experienced a protection display by the women of a village near the sacred mountain of Ben Bulben in Co. Sligo, Ireland, after the farmer husband of one of them had threatened to bulldoze the whole setting. When the bulldozer arrived the community's womenfolk all joined arms and encircled the thorn wrath to protect it. The Hawthorn was itself decorated with ribbons and garlands in honour of its importance. This was called 'bawming'. It is as if the sacred mound was garlanded in veneration.

There are extraordinary legends from ancient Greece about humans being turned into Cypresses. The Cypress became the symbol of mourning and inconsolable grief. Lucan the Roman poet (AD 39-65) records this. Even Virgil (70 BC) uses Cypresses in the funeral rituals his heroes

A princess or queen from the royal family of the Sumerians wears a gold version of a flower garland.

The garlands or wreaths worn by the royalty of the Sumerians or Assyrians, from as far back as 2000-3000 BC, demonstrate that this convention is archetypal, and most likely represents the 'flowering' of what is called the crown chakra by the Hindus.

Cypress trees have always been found on
sacred sites and hold a special place in
the human psyche, linking the lower and
upper worlds.

were subject to. In England as recently as 1889 people made Cypress garlands at funerals. It is also a recorded habit for some Muslims to plant a
Cypress over a grave site.

Oak garlands, home woven and never carried over the threshold, were
displayed in parts of Germany in windows or on front doors to dispel the
negative influences of witches at certain times of the year. The meaning
of the witches in the opening scene of Shakespeare's *Macbeth* has been
well explained by G. H. Mees where he proposes that they symbolize the
'fates' of past, present and future.

An Olive wreath (as distinct from the Olive branch of peace) was
sometimes given as the highest mark of recognition to winners of the
Olympic and other games, also as a reward for military merit. In Japan
the Olive tree is still the symbol of success in study and in civilian and
military matters, being considered the tree of triumph. In Christianity
the Olive is associated with the Virgin Mary. St John Chrysostom says,
'The Virgin Mary is the beautiful Olive tree, which has yielded the
blessed fruit of majesty'. The oil symbolizes 'benevolence and alms'.

The same oil is symbolically taken as 'sacred' and is used for 'eternal'
lighting. The fuel in sanctuary lamps was invariably Olive oil with some
beeswax added. The same was true for the sanctuaries of Athena, Daphne
in Delphi, as well as in the ancient Vedic tradition. Even the caduceus of
Hermes or Mercury was said to be an Olive branch with two symbolic
snakes wrapped around it. (Mercury is Hermes or Thoth, the god of communication or messenger of the gods, stemming from ancient Egypt.)

The image of the Pomegranate is invariably stylized in its representations, particularly in Europe. It has become the emblem of both Iran and Spain. The German Emperor Maximilian I (1459–1519) used the same emblem, despite the fact that the fruit bears no particular beauty or perfume, though it does have an abundance of seeds. Thus it was naturally a symbol of fertility. In Iran or Persia followers of Zoroaster/Zarathustra (c. 630 – c. 550 BC) made garlands of Pomegranate twigs to provide protection from demons and negative forces.

In ancient Greek mythology Heracles/Hercules, the largest and most masculine of the heroes, decided to adorn himself with a garland made from the leaves of the White Poplar which lit up his forehead. Adorned like this he was able to find his return path from the underworld. It was the sweat of Hercules' brow that turned the underside of the leaves white. Followers of the cult of Hercules all wore garlands of similar White Poplar branches as symbols of courage and heroism. Youths who won fighting contests were rewarded with such garlands. The white wood of the Poplar was the only one considered appropriate to be used for offerings to the supreme God Zeus. When opened, Sumerian graves dating from 4000 BC revealed golden wreaths or crowns depicting White Poplar leaves.

We also have the Laurel wreath that was placed on the heads of heroes in ancient Athens. This relatively common plant symbolized the highest status a human may reach: success, victory and glory. It is dedicated to Apollo (a name meaning 'one, not many'). It gives good fortune and confidence. It is the plant of the sun mythologically and astrologically. ('Astrology' is used here with its original meaning of 'logic of stars'.) The

The three 'weird' sisters or witches in Macbeth are the fates that determine the soul's fortune: one for the past, one for the present and one for the future. The cauldron is the melting pot for all souls.

The Pomegranate fruit was highly valued in antiquity for its abundance of seeds. A natural symbol of fertility, it was adopted by more than one country as national emblem.

The ancient and universal tradition of wearing garlands to beautify and show respect is still very much alive. Above, a Ukrainian girl at a festival; above right, a girl from London wears a Rose garland for her celebration; and below, the African girl wears her version of a celebratory garland with added face colouring. Flower garlands lend dignity as well as beauty.

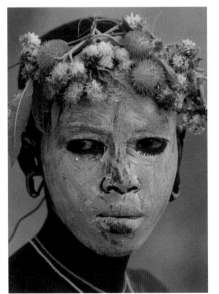

symbolism of the Laurel wreath can be traced back into Greek mythology. When Daphne, an attendant of Athene the goddess of wisdom and skill (and even conflict) was being pursued by Apollo, she prayed as she fled that the gods would cause the earth to open and swallow her up — or else change her form to distract the pursuer. Her prayers were answered at the very point of being caught by Apollo. Athene transformed her into a Laurel tree. This is why it is called *Daphne laureola*. Apollo then decided this Laurel tree would be his favourite and personal tree. It is an 'evergreen' with intoxicating properties. As a wreath, it also symbolizes poetic inspiration and timeless fame. We have all heard the phrase 'resting on one's laurels', indicating a form of laziness — or resting awaiting inspiration.

Victors at the Pythian games held at Delphi since the seventh century BC in honour of Apollo, the slayer of the python, were crowned with Laurel leaves having achieved in music, poetry, painting, sculpture athletic sports, chariot and horse racing. Even orators and philosophers were so honoured right through until Roman times. The metaphor is still much alive, in our use of the word 'laureate'.

In *The Knight's Tale* Geoffrey Chaucer describes how Emelye ('that fairer was to sene / Than is the lylie upon his stalke grene') makes a subtle garland of red and white flowers for her head while performing her 'observaunces' to May in an enclosed garden. 'She gadereth floures, party white and rede, / To make a subtil garland for hire hede.' This miniature is from a manuscript of Chaucer's source story, *Il Teseide* by Boccaccio, illuminated by the Master of the Hours of the Duke of Burgundy around 1465.

One of the most significant of the religious paintings of Albrecht Dürer. It
was painted for a Christian community in Venice on his visit to Italy. Note the
garlands being donated to the deserving.

Roses and Wreaths

The rose is a great deal more
Than a blushing apology for the thorn.
Rabindranath Tagore

Crowning with a wreath of Roses held a very special meaning. Aphrodite and Eros both wore wreaths of Roses in the times of ancient Greece and Rome. The three graces, the hand maidens to Aphrodite/Venus, also wore the same. Hebe/Juventas, the Roman goddess of youth, Ganymede who was taken from eternal youth, and Erato the muse of erotic poetry and hymns, all wore Rose wreaths. There were three more Rose-wreath wearers in the Greek/Roman pantheon: Erato's sister Thalia the muse of comedy, Pax goddess of peace, and Comus the god of banquets. At a higher level Dionysus/Bacchus was occasionally depicted wearing a wreath of Roses. Artistic expression places the Rose as symbolic of love, festive pleasure (most crowns of joy are made of Roses), beauty and martyrdom. Thus the outer garland expresses the inner state of being and all can be included in the deeper mythology of flowers.

England's own Queen Mary Tudor (1516–1558) was portrayed by the artist Anthonius Mor (c. 1520 – c. 1577) holding a red Rose, in a painting which is now in the Prado Museum in Madrid. In the Christian tradition we find Roses particularly represented as integral with the Virgin Mary. Mary often holds Roses and Lilies and scatters Rose petals over the martyrs. In Renaissance paintings a garland of Roses is often a reference to the rosary (the prayer beads) of the Virgin Mary. The Rose as a symbol of Christ is also clear.

There is yet another subtle symbol of the Rose as a 'receiving cup' that clearly indicates the Holy Grail. Even the five wounds of Christ are sometimes represented as Roses. We must not forget the name which the great cosmological stained-glass windows were given in European cathedrals: 'Great Circular Rose' windows.[22] Those at Chartres appear to the author to be supreme of their kind — symbolizing visions of all time.

In the Christian tradition paradise is represented as a multifoliate Rose, particularly by Dante to whom we will return in the final chapter. Mary is often compared to a mystic Rose as we have already noted. The

This special garland of red and white Roses was made by the contemporary Master Boon Pan from Thailand, now working in London for 'Tiger Lily Florists', Balham.

Here we see Queen Mary Tudor in a portrait by A. Mor, the Dutch artist at the Spanish court. Symbolically, she holds a red Rose in her right hand — meaning far more than a casual affection for Roses.

The white Rose has a natural purity as well as a national symbolic value for those who live in the United Kingdom.

The red Rose has been called a symbol of the passion of Christ in the Christian world. The perfume of the Rose is considered the most spiritually sublime by the Islamic Sufi fraternity.

white Rose symbolizes virginal innocence and purity. The red Rose became symbolic of the blood of Jesus, symbolizing the highest expression of passion. In medieval Europe the Rose was a common symbol of love, beauty, happiness and silence. (This sacred aspect of silence is closely afforded to all flowers when there is no breeze. Silence being the way to God.) But with more vernacular symbology they could represent sensuality. 'Sub Rosa' or 'not to be repeated beyond this room' refers to the plaster rose in the ceiling of a meeting room. This symbol of silence was introduced into England in the seventeenth century, and was used in many private or secret society gatherings.

The Rose can also be taken to represent the totality of the wheel of passing time. The fivefold petal pattern is taken to refer to the five elements and the recurring cycles of the cosmos. Aristotle spoke of this cycle of the five elements. In ancient times it was customary to wear a Rose garland in the spring symbolizing the eternal rebirth after the 'death' of winter. Hecate, goddess of the underworld, is sometimes represented wearing a crown of Roses with five petals. In their monumental reference book *Compendium of Symbolic and Ritual Plants in Europe*, M. de Cleene and M. C. Lejeune[23] talk about a seven-petalled Rose, which they relate to the seven directions of space, the seven planets, seven days of the week or seven degrees of perfection. (The directions of space are usually depicted as radiating from a 'still' centre in six directions, which can be taken as a representation of a three-dimensional cross.) Such simple assertions do require a little more explanation as natural Roses do not have seven petals except in extraordinary circumstances. However the construction of the Rose in this case is clearly for symbolic purposes, the sacredness of seven being united to the sacredness of the Rose.

The Rose must be near the summit of natural flowers, particularly for its sublime perfume. It is celebrated by all civilizations and is recorded as having been cultivated as long as 5,000 years ago in the ancient gardens of western Asia and north-eastern Africa. It figures in the Judaeo-Christian Bible as the Rose of Sharon in the *Songs of Solomon*. Sa'dī, the great Persian poet of the *Gulistan* in the thirteenth century placed Roses in pride of place in his inspired poem. Roses were cultivated in the mythical gardens of both Semiramis Queen of Assyria and Midas King of Phrygia.

There is also a moving story of Chloris, the goddess of flowers. One cloudy morning as she walked through the woods she found the body of a beautiful nymph lying in her path. Saddened to see such a lovely creature dead she decided to give new life to her by transforming her into a beautiful flower surpassing all others in charm and beauty. She called on the other deities to help her: Aphrodite to give beauty, the Three Graces to bestow brilliance, joy and charm, and her husband Zephyrus, the West wind, to blow away the clouds so Apollo could shine and offer his blessing as the sun. Then Dyonisus, the deity of the grape vine, added nectar and fragrance. When the new flower was finished, the gods rejoiced over its charming beauty and delicate perfume. Chloris collected a diadem of dewdrops and crowned the new flower as the 'queen' of all flowers. Aphrodite then presented the Rose to her son, Eros the god of love. Thus the white Rose became the symbol of charm and innocence, and the red Rose of love and desire.

Three of the most grand, beautiful and significant of the rose windows to be found in European cathedrals are at Chartres, one in the south as shown here, one in the north and one in the west front. All three are twelvefold. They all glow with the genius of medieval coloured glass, itself a mystery. They are treasures of European Christianity with immense symbolic content and meaning.

Sathya Sai, the contemporary spiritual spokesman for the Hindu faith in India, encouraged his sevenfold yantra to be published. It is available in traditional metal form for aiding meditation.

The Inspirational Perspective

Unity is possible through the contemplation of geometrical figures, diagrams and symbols (analogous to the Hindu Yantras and Mandalas) projected into the Imagination which occupies, according to Proclus, the central position in the scale of knowing.

Algis Uždavinys

To see a World in a Grain of Sand
And a Heaven in a Wild Flower . . .

William Blake, from Auguries of Innocence

Lily of the Valley

No, it is not different,
Now I am old,
The meaning and promise
Of a fragrance that told
Of love to come
To the young and beautiful:
Still it tells
The unageing soul
All that heart desires
For ever is
Its own bliss.

Kathleen Raine

The great astronomer Copernicus wished to be painted with the Lily of the Valley, his inspiring flower.

Similarly, the influential European philosopher Erasmus wished Albrecht Dürer to place his inspiring aromatic Lily of the Valley immediately next to him.

The highest level of 'knowing' we propose is the inspirational level. Needless to say flowers have a long history as a source of inspiration for human beings. The word 'inspiration' has, like so many English words, suffered with regard to its meaning over the years. The original meaning links it to the breath (*pneuma*) as symbolic of soul or spirit. Inspiration is thus similar to the taking in of a full breath. Thus once the lungs are full of breath metaphorically the body is filled with spirit (*pneuma*). The result is often also associated with *enthusiasm* which originally meant 'god-filled' in Greek. Plato, through Socrates, taught that there were at least four inspirations; these were called 'manias'. Socrates said that there were two types of *mania* or madness, one arising from human disease, the other *when heaven sets us free from established convention*. This is the source of the saying: 'mad genius' and is beautifully described by

Clio   Thalia   Erato   Euterpe   Polyhymnia Calliope Terpsichore   Urania   Melpomene

The nine muses (Apollo being the tenth) with the instruments
of inspiration for the classical world.

Apollo sits among the inspiring muses.

Socrates in the *Phaedrus* dialogue by Plato. Socrates speaks:

> And in the divine kind we distinguish four types ascribing
> them to four gods. The inspiration of the prophet to
> Apollo; that of the mystic to Dionysus; that of the poet
> to the Muses and the fourth type, which we declared
> to be the highest, which is the madness of the lover, to
> Aphrodite and Eros.[24]

So we can see a similar fourfoldness in the inspiring forces as is inherent in our 'levels of knowing', as we might call it. The scale is as follows:

Eros as the inspiring God of the Amatory Inspiration.
Apollo as the inspiring God of the Prophetic Inspiration.
Dionysus as the inspiring God of the Mystic Inspiration.
The Muses as the inspiring Gods of the Poetic Creations.

Bacchus or the god of the vine and wine by Michelangelo. Another source of inspiration yet 'wine' holds a symbolic meaning beyond mere alcohol.

Plato is clear in the *Timaeus* about the nature of inspiration, enthusiasm, 'mania' and so forth. He states that to be gripped by an inspiration or enthusiasm one is in '*quite a different state of mind than normal consciousness*'.

There is an excellent and enlightening series of commentaries on Plato's works by a group who called themselves the 'Scholia of Hermeas'. Thomas Taylor, the exceptional eighteenth-century English translator and interpreter gives us most helpful translations of their work — particularly on the inspirations. They said in their comments of the *Phaedrus* dialogue of Plato:

> Indeed it is obvious that into any production of any of the
> four arts, all the four inspirations must enter else it cannot
> be perfected . . . Inspired art is a sacred thing, not merely
> a means of sensual or emotional delight.

The four inspirations and four timeless values are obviously qualities rather than things. The first we will call Truth. When the truth of an issue becomes clear then it normally inspires action or deep contemplation; it is also the highest standard which we are inspired to work to. Each flower is truly itself.

The next is the Good (which has a special value and meaning to the Platonic School) or one could say Goodness as this will always raise the human spirits and will inevitably be its own reward. To discover that the activity that one is engaged in has 'come good' will always inspire. A good result to any effort holds its own inspiration and is the very springboard for the next effort or task. All flowers are good, in their own terms. For us as well as for themselves and their environment — as long as we avoid eating the poisonous ones.

The third, we suggest, is the Beautiful. This is the hardest to describe in words as it is normally a state of being or consciousness. Plotinus consistently advised that we should always seek *beyond* the beautiful experience to reach 'Beauty Itself'. Beauty is possibly the greatest and the most subtle of the inspiring qualities — as well as the most mysterious, like life itself. How much do we really know about the inner meanings of flowers and their unquestionable beauty, and how able are we to put this into words?

The fourth value, we suggest, is Harmony which is the integration of all of the above. This is again one that is difficult to put into words but is surely a great inspirer when experienced, as it brings together in accord all factors and qualities under consideration. All flowers emanate harmony naturally.

One of the most subtly simple yet treasured Chinese plant paintings, attributed to Cheng Ssu-hsiao. Its 'value' to the emperors and owners is marked by the number of ownership seals around it.

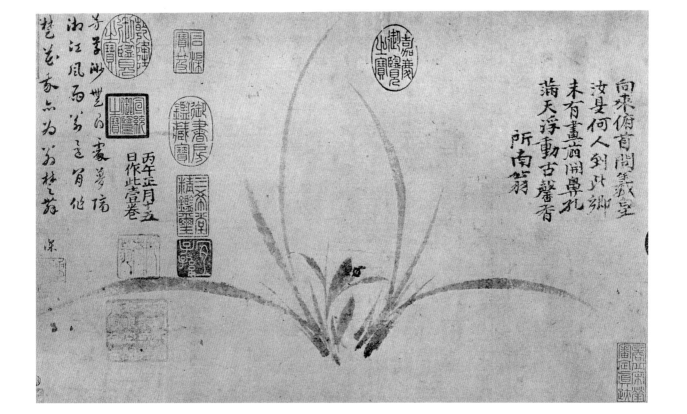

Harmony can be a relationship between two or more things or ideas, it can be a musical experience, and it can be a sublime inner experience in natural surroundings. It is often associated with great peacefulness, so becoming the inspiration of calmness. When the Soul is in harmony with itself we call it 'peace of mind'. Flowers are natural promoters of harmony for us and certainly nurture 'peace of mind'. This is clearly evident in the hospital ward environment.

Flowers have inspired painters throughout recorded time. China and Japan may offer the greatest number, not forgetting the miniatures of Islam. In the West, the most striking examples include Leonardo da Vinci and Albrecht Dürer. One of the most renowned European creators of flower paintings in the late eighteenth and early nineteenth centuries was Pierre-Joseph Redouté. In contrast to the earlier artists, there is no philosophical or mythological depth to Redouté's work although his pictures are 'sound' and representationally faithful to the originals. This loss of the symbolic in representational art eventually became the cause of the invention of the photographic camera.

Another specifically sensitive plant study by Leonardo da Vinci.

The greatest need for inspired action in our times has arisen from negative, usually mechanical, forces universally at work and reflecting a contemporary breakdown in our relationship with Nature — mostly stimulated, unfortunately, by unbridled financial greed.

The rediscovery or recovery of what is natural is a matter of great urgency. The inspiration may come from the urgent inner need to experience *the truth*, the *good* in itself as well as in pure *beauty*. Flowers have always been available to help in this urgent need for a new inspirational impulse to live naturally and peacefully, in harmony with the world.

If it is taken as natural (as Socrates did) that there are four levels to existence then the most helpful understanding of these 'levels' or 'layers' of meaning is to read them as inherent and intrinsic to our human nature. If we are, as is held, a microcosm then we have literally everything within us. As Albert Einstein used to insist, the most remarkable aspect of the cosmos is that we can understand it, however modestly. (Although at the time of writing it has been claimed that 'we' only know about five per cent of what is in our universe — which only goes to highlight what meaning we give to 'know'.)

This inspiration called the 'prophetic' relates to discursive reasoning which in the Socratic tradition is named *Dianoia*. Apollo is called the leader, not only of the nine muses, but also of *indivisible harmony* and intellect. Whereas the lesser inspirations find truths with a small 't', the prophetic under Apollo's power of unity finds Truth with a capital 'T'. This was also described by the Platonic succession as raising the inspired soul into the eternal 'now'. This inspiration enables the soul to realize the essential oneness embracing all differences. (It is of special interest that in *The Golden Verses of the Pythagoreans* we learn in sentence xxiv: 'And thou shalt know that Law hath established the inner nature of all things alike'. Commentary by Hierocles of Alexandria.)

Next we enter the all-embracing inspiration which may be called the 'Amatory' or the inspiration of universal love. Eros, the 'first born' of the gods presides over this primary inspiration. The love of the True, the Good and the Beautiful is concluded and fulfilled in this 'contemplation of reality itself'. This at-one-ment (with the godhead) is called *noesis* in the Platonic tradition. This consummation of all inspirations comes under the heading of love. Proclus, the last and greatest of the Platonic commentators, explains in his 'theology' the fuller dimensions of the inspiration of love. Eros is described as he 'who first leapt forth from [the Divine] Mind, clothing himself with the fire with which he is bound'. It hardly needs reminding that a particularly inspiring utterance by Jesus the Christ was 'Love thy neighbour as thyself', which has prompted the question, 'what isn't my neighbour?'

Proclus explains: 'Love converts all things and assimilates them into the nature of the Beautiful. Love, therefore, supernally descends from the Intelligibles into mundane natures, calling all things upward to Divine Beauty' — or *Beauty Itself* as Plotinus would have it. This is the transcendent role of flowers.

Proclus said that the first 'idea' of the beautiful is in the intelligible Divine intellect. 'This beauty is of a vital intellectual form, the source of *symmetry* in all things'. He explains that there are three Hypostases in the Intelligible Gods: 'these three . . . unfold themselves into light in the ineffable order of the Gods as Faith, Truth, and Love'. Love being considered the ultimate inspiration.[25]

In these clarifying commentaries on the sources of the Western tradition we can detect the implied journey: 'Beauty itself descending, as it were, to earthy existence eventually becomes integral, to a degree, with all things and still "gleams" in natural objects, not least flowers, so the soul, inspired by love, may ascend from the sensible beauty of outer forms through moral, intellectual, spiritual and eventually external forms of Beauty to arrive at contemplation of the beautiful itself'. (From Thomas Taylor's translation of the *Scholia Hermeas*.) The language is symbolic but the implied meanings are not difficult to interpret.

All the inspirations are in need of each other. The prophetic 'mania', as Plato calls it, uses the musical inspiration spontaneously. Yet it is clear that all levels of inspiration can be 'read' or 'learned' from the flowering

An intense study of grasses by Albrecht Dürer. The Jewish Bible or 'Old Testament' declares significantly that 'all flesh is grass': a truth as scientific as it is biblical. Our physical bodies are built on the green world either directly or indirectly.

world — at best in silent contemplation. In flowers it is the colours that 'sing' as well as the murmuring of the innumerable bees.

What more need be said of the relationship between the inspirations and the floral world? Flowers, the embodiment of beauty, feed the developing intellect by their symmetries, colours, geometries and forms. They hold much in common with musical proportions as the same archetypal number pattern informs both. The soul's very nature was proportion according to Plato's *Timaeus*.[26]

Having asserted that there is a proportional or numerical relationship between music and the geometry of growth patterns we will now look at what is called the Fibonacci sequence, the one that leads towards the golden proportion. This can be represented by a logical series of numbers: 1, 1, 2, 3, 5, 8, 13, 21, etc. This sequence is created by starting from 1. As there is no other number yet to add to this 1 another 1 is placed second. Now by adding the last two numbers (1+1) we arrive at the third in the sequence which is 2. Next we add 2 to its previous 1 to get 3, then 3+2 to get 5, and so on. If we consider this number sequence which is powerfully evident in the unfolding of the vegetative world, we can see the analogy with musical notes which sound beautiful to the ear.

To explain this sequence, firstly we take 1 as the leading note. Then 2:1, the proportion of the octave. Next 3 notes or the first 'chord'. Followed by 5, the pentatonic scale, 8 the diatonic scale, 13 the full chromatic scale, 21 the Indian *Shruti* scale. That the human musical scales are in the same numerical series surely indicates some sort of universal harmony between number, forms, proportions and sounds — particularly as we experience them all as 'beautiful' in their own sphere.

Eros is seen racing from the scene of his crime (trying to steal the honey of the bees). An allegory illustrated by Albrecht Dürer, signifying that there is a sting in the tail of stolen love.

The unfolding of this golden series of numbers recurs quite evidently in the floral domain. Hence this series of whole numbers relates not only to geometry and music but also to the flower world. This is yet another example of how numbers and proportion in both the flower and the musical world harmonize.

When we see an inspiring image we are immediately prompted to communicate it. However the 'language' that we choose to communicate with will naturally 'colour' what is received by another. So when considering how to best describe or communicate anything it is well to remember the following about language conventions:

The language of the senses suits the senses.
The language of feelings suits the feelings.
The language of thoughts suits thought.
The language of a culture suits that culture.
The language of spirit suits spirit.

Thus when expressing yourself be careful which language you use for which area, as they are not automatically transposable.

Further:
The language of numbers suits numbers.
The language of geometry suits the geometrical.
The language of harmony is best heard or seen as harmony.
The language of the cosmos suits the cosmos.
We can equally say that:
The language of flowers is beauty:
Beauty of form and symmetry,
Beauty of colour contrasts,
Beauty of unfolding movement,
Beauty of fragrances,
Beauty of movement in the breeze,
Beauty of setting, natural or human made,
Beauty of the many different greens.

Language is for communication.
Beauty communicates most immediately and
always in its own language.

In short, communication is *about* something and is not necessarily the thing itself. Although it *can* be the experience itself as in poetry, music and meditation.

The intensity of colour as well as central symmetry make these flowers powerful
examples of the attractiveness of such intensely saturated blooms.

Number and Flowers

*Numbers then hold the keys to understanding the
organization of the world.*

D. J. O'Meara, from *Pythagoras Revived*

We are so habitually used to grasping the simple number groups immediately that we rarely pause to consider what they might mean to us at a
deeper level of consciousness. Each flower is representative of a number
either in its wholeness, or in the number of petals it displays. This is also
true of the leaves and their sets, as well as the sepals and the flower head
buds coming from the same stalk.

It is well to remember that the ancient Greeks did not invent separate
symbols to represent numbers. They used their alphabet, and handled
pebbles — these later became the *calcis* in Latin from which we get
'calculation'.

Each number represents a universal law and possesses its own
individuality, its own character and thereby its own meaning to the
human consciousness on several symbolic levels — although today,
sadly, they have been reduced to quantity alone.[27]

No civilization has ever neglected the qualitative meaning of numbers. For Plato numbers led to an appreciation of 'the highest level of
knowledge' both inwardly as well as outwardly. Proportion (and harmony) between numbers was seen by him as of the nature of the soul,
both cosmic and human.

The ancient Chinese founded their civilization on the knowledge
of the value of what can be called numerology in its traditional and
authentic sense. (Numerology is a much maligned word which actually
means the logic of numbers.) The correct understanding of numbers
links to both micro-cosmos and macro-cosmos. Albert Einstein said that
it is a wonder that we can know of the universe through number. The
understanding of number established cultural, social, and civilizational
harmony. Both Iamblichus (*c.* 250 – *c.* 330) and later Nicholas of Cusa
(1401–64) advocated the study of numbers as they believed it was
the nearest the human mind could get to the Divine Intellect. In fact
the later Platonists described arithmology as a 'participation' with the
Divine Mind. This Platonic/Pythagorean wisdom tradition consequently

As 5 is to 8, 8 is to 13 and 13 is to 21 in
the golden sequence. Here the Daisy
type of flower radiates its thirteen petals
beautifully.

SECTION OF PIANO KEYBOARD
current international convention

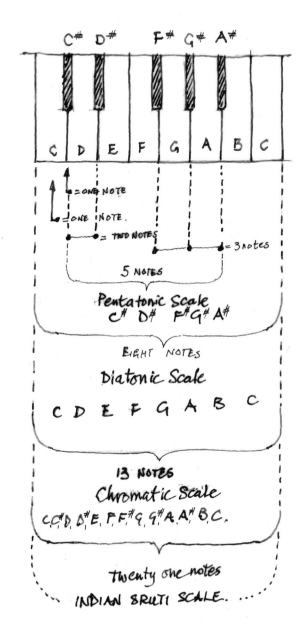

1 = C or C♯

2 = C, D or C♯, D♯

3 = F♯, G♯, A♯

5 = Pentatonic scale
      C♯, D♯, F♯, G♯, A♯
      Ancient China

8 = Diatonic scale
      C, D, E, F, G, A, B, C
      Christian & European

13 = Chromatic scale
       C, C♯, D, D♯, E, F, F♯, G, G♯, A, A♯, B, C
       Arabic & Islamic

21 = Indian Sruti scale
       Vedanta and Islamic

The scrutiny of our ear's sense of beauty reveals that the golden proportions
arise in the numbers and relationships between the notes — both mathematically
and culturally. Most folk music is pentatonic.

influenced the development of science in Judaism, Christianity and Islam simultaneously.

Numbers were considered to be the mediators *par excellence* by such as de Saint-Martin, not only governing the laws of space, time, life and substance, but also offering access to the First Cause itself. 'Numbers are principles', he said, that are 'coeternal with Truth'.

In the higher sense even geometry was understood to be the basis of harmony between forms, as it was considered to be 'number in space', quite transcending physical earth measure in its intelligible origins. Angelus Silesius put it this way: 'God is in all things as unity is in all numbers'.

The Aztecs said without numbers there was no cosmos. For them each number was associated with a particular god, a colour, a direction, as well as groups of good and evil influences.

The Pythagoreans held number in the highest reverence in their oral doctrines, which apparently were never committed to writing. These doctrines can be taken to be reflected in the inner truths of Plato's *Timaeus* dialogue, as Timaeus himself is understood to have been a Pythagorean. Hierocles the fifth-century Alexandrian Pythagorean tells us so explicitly in his commentaries on *The Golden Verses of the Pythagoreans*.

The fundamental symmetries available in the nature of our universe is relatively small. Here we see one we are most familiar with 2x2 or 4. This fourness characterized so much of our experience of our world. From four directions to four states of matter or the elements of earth, air, fire and water. Four levels of awareness are found in the human psychology according to Socrates and the four gospels are fundamental to any Christian.

This Chinese zodiac demonstrates the universality of astronomy or study of the patterns of the heavens. Here we have layers of symbolic as well as literal meaning.

Number and Numbers:
Quality As Well As Quantity

Pythagorean and Platonic mathematics deals with realities that are intermediary between (1) immaterial and indivisible intelligibles and (2) material and divisible sensibles.

Algis Uždavinys

Numbers are universally true to all people by nature. If we wish to count or even measure to any extent we cannot do so without numbers. Numbers offer us both objective and consensus order. Number and numbers are intrinsic to our nature from the micro level of molecules to the details of our limbs and to our ability to do mental arithmetic. Each number has a qualitative as well as a quantitative value. In other words, numbers are as much living symbols to human kind as they are reliable and objective tools for calculations. Plato quite specifically introduced 'Oneness' to emphasize the quality of Unity above and beyond the number 1 alone.

We may say 'two of something' but when we say *twoness* we mean a great deal more. The first statement stays where it is. The second (the quality of twoness) is unlimited in its symbolic possibilities. The first could be described as A=A, the second as A+ any other letter of the alphabet; which would be vast in its implications. Symbols imply *correspondences* and it is these that ultimately cohere the universe. It is the profound nature of relationships which cohere our universe. Coherence is relationship.

In this sense the numbers of petals that a flower displays is bound to have resonances with our deeper consciousness. In this way we can appreciate how a flower might indicate 'meaning' in the abstract sense. Threeness in a flower is not less than a reflection of threeness itself. How we choose to *interpret* that threeness is up to ourselves and our quality of imagination, our poetic genius as well as our practical sense — and also according to the conventions of our civilization or culture.

From the geometrical view point the numbers are quite specific in terms of their balance of symmetry and shape.

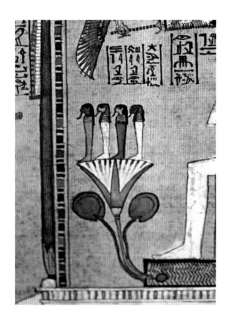

In Egyptian symbolism the four children of Horus (Imsety, Hapy, Duamutef and Qebehsenuef) are the guardians of the four cardinal directions. They are also called 'The Sovereign Princes of Osiris'. Here we see them being born out of the primal Lotus.

Oneness

Twoness

Threeness

Fourness

Fiveness

Pebbles as the integral number patterns
in geometrical relations.

Threeness is the first possible 'shape' when it comes to straight line polygons. In other words, it takes a minimum of three straight lines to make a triangle. We do have 'twoness' in the symmetry of most leaves (the handedness of their sides) in the same way that we have a twoness in flower petals. Twoness or duality symbolizes another primary law in our experience — that of sameness and otherness or more personally 'me' and 'all others'. Thus twoness can be a profound study as well as a fundamental experience. Our heart beat is twofold: systole and diastole. There is virtually a polarity in all our bodily functions, from blinking to swallowing and to breathing. However it must also be recognized that twoness is both divisive and the seat of all conflicts and oppositions unless seen as a complementarity. 'This', 'that' and the 'conjunctive', the latter being intangible yet the most important — are epitomized in balance between Yin and Yang and the containing Tao.

One, another, and that which links them as *relationship* is the first principle of recognizing order. Order *is* relationship. Relationship is the intangible link between one thing and another, drawing them together into a harmony. That which creates relationship can be said to be the beginnings of all cosmic order. The better the qualities of social relationships the better the society is as a general rule. The more harmonious and well-related that society is, the more 'civilized' it becomes. Threeness and harmony are always linked, as *harmony* itself is the finest relationship that can be nourished between two entities. It both honours their differences yet binds them into one.

One establishes, two divides and doubts, three reconciles. To summarize the quality of three we have the translation of Du Bartas made by Joshua Sylvester in 1578,

> The eldest of odds, Gods, number properly . . .
> Heaven's dearest number, whose enclosed centre
> Doth equally from both extremes extend,
> The first that hath a beginning, mid, and end.

As we have already noted, some philosophers have called *three* the first number — because One is the *principle* of unity and sum, Two the *principle* of multiplicity — so Three is the first *individual* number.

Learned studies have been published on the universal profundity of threeness or trinity so we will have to leave it to the naturally inquisitive and serious seeker to journey on. Threeness in the plant world is clearly manifest in root, stalk and flower; also in the fruit as husk, flesh and kernel. There are few if any religions that do not have a trinitarian aspect to their deity or deities. Triangulation has always played a primary role in human ritual and symbology — even to the level of Divinity itself. The physical triangle made of minimal struts is the law of structure in all physical engineering.

Fourness (often regarded as a double twoness) so clear in the Clematis and the Wallflower, represents the four directions and elements of the material world. The *four* obvious phases of time defined by the moon (intimately connected with all biological rhythms) are evident in crescent, waxing, full and waning. Fourness reflects the spatial directions, the seasons, the phases of life and so on. So it is in *fiveness* that we find the golden ratio, particularly in the arms of the regular five-pointed star or pentacle. Fiveness is also the centre of fourness.

Fiveness pervades the qualities of life in so many traditions, not least the ancient Chinese. Five elements, five sounds, five flavours, five types of grain, five sacred mountains and so on. Ancient Chinese civilization is built pentagonally.

John Donne's witty poem *The Primrose* sums the relation to 'fiveness':

> Live Primrose then, and thrive
> With thy true number, five;
> And women, whom this flower doth represent,
> With this mysterious number be content;
> Ten is farthest number; if half ten
> Belong unto each woman, then
> Each woman may take half us men;
> Or if this will not serve their turn, since all
> Numbers are odd, or even, and they fall
> First into this five, women may take us all.

From fiveness we naturally move to sixness, the number of Crocuses, Tulips, Daffodils and Lilies. Six is the number of the crystalline world as well as the 'intervals' of creation equally for Jews, Muslims and Christians. The Lily seems to be a sixness but is more correctly two threenesses. Six was also called a perfect number as it is the sum of its divisors in two quite different ways: 1+2+3=6 but also surprisingly 1x2x3=6. Those who have read deeply into this sixness are of the highest calibre in the Abrahamic faiths. And the most ancient Chinese book of wisdom and prophecy, *The Book of Changes*, was structured on the six-layered hexagrams — that is, hexagram readings with six levels of symbolism. The graphic representation of sixness as geometrical symbolizes both breadth and depth in Judaism and Vedic India. The geometrical hexagonal home

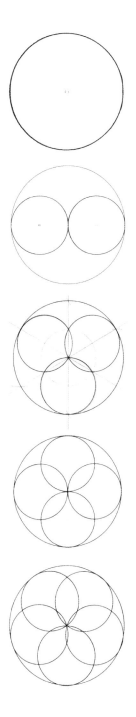

If we take a circle and draw two, three, four and five circles of half radius within it, we get the following geometrical flower forms.

Left. The Iris flower is a classical threeness.

Centre. The Clematis has many symmetries. Clearly a fourness here.

Right. Fiveness predominates in the primrose family but we can find sixes and even sevens too.

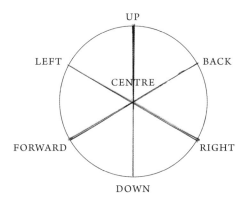

The directions of space when made 'human' relate to the basic directions we can act or move from a still starting position.

of the bee has been another representation of Creation to some, as the cells demonstrate the perfection of economical use of their wax.

Sevenness is rare in flower geometry — although not unknown. Sevenness, among the shapes between one and ten, is the most mysterious of number qualities and was revered particularly by the Pythagoreans. This was another reason why it was called the *virgin* number, associated equally with Athena who sprang from the head of Zeus and the Virgin Mary as the mother of Jesus.

Eightness is inevitably associated with the octave and anciently the seven visible planets were completed as an octave when viewed from the earth. The seven notes within each octave was the basis of the concept of the 'music of the spheres'. Eightness is sometimes called the auspicious number: it is a form of perfection of the earthly order, giving the four cardinal directions and their intermediates. The ear easily registers the eighth note in the diatonic scale as the octave. Eightness is part of the Fibonacci sequence as we saw, so it is likely to arise in the flower world — which it does in particular in the beautiful flower known as Cosmos as well as in the Clematis and Lotus (to which we will be returning).

Eightness is associated with paradise. The word 'paradise' is of Iranian origin and means 'the walled garden', a clear reference to humankind's role as the gardener and cultivator of the green world of flowers. In ancient Persia gardens were designed and divided into an eightfold pattern. This pattern also spread to India. We find this symbolism best represented in Sa'di's *Gulistan* or 'Rose Garden' which is divided into eight chapters. Not only is there a reference to eight paradises but Islamic mythology offers the model of there being eight angels holding up the throne of God. (From a *Hadith* of the Prophet Mohammed. A *Hadith* is a traditional saying recorded as being spoken by the Prophet.) Spiritual expressions of this number occur widely, the best known probably being the noble eightfold path of Buddhism. Maybe there is a connection between the vertical 8 and the horizontal sign of infinity?

There were eight 'immortals' in ancient Chinese mythology one of whom, Ho Hsien-ku, held the magic Lotus flower, which has eight petals. Further supporting the domain of flowers was Lan Ts'ai Ho who carried a whole basket of flowers. For Confucius there were eight precious items — just as there were eight musical instruments in ancient times.

Nineness rarely occurs in the floral displays of petals but definitely in some Cactus plants. Nineness is best seen positively as 3x3 or a trinity of threenesses. In Christianity we find nine orders of angels cited by Dyonisius the Areopagite. Nineness is in evidence in some ancient Chinese Buddhist structures, for example pagodas with nine stories. These were related to a ninefold division of the heavens. Nineness pervades the structure of the platform of the Temple of Heaven in Beijing. The quality of nines also pervaded the ancient Egyptian theology.

Another example of nineness ('the nine states of existence') is found in the seminal encyclopaedia *The Rasa'il*, written by the Brethren of Purity in tenth-century Iraq. This work is quite urgently needing a complete English translation although the leading Islamic philosopher Dr S. H. Nasr has given an excellent introduction to the depth of this work.

Shakespeare uses nine in a particular way in *Macbeth*. The scene is on a stormy heath where the three witches have arranged to meet Macbeth. They chant:

> Thrice to thine, and thrice to mine,
> And thrice again, to make up nine
> Peace! The charm's wound up.
> *Macbeth* Act I scene III

These enigmatic three lines offer quite significant analysis for the truly curious, not least indicating how a charm is 'wound up', no doubt to its maximum power.

Apollo was accompanied by nine muses. Plotinus the illuminated Platonic philosopher wrote his most influential work in the books known as the *Enneads* (each containing nine treatises). Nineness can be traditionally regarded as a completion of the independent archetypal numbers, as ten is the return to Unity. This return is called the Pythagorean Tetraktys, based on 1+2+3+4 equalling 10.

However tenness is yet another mysterious expression of the golden ratio and finds most significant expression in certain rare geometric

The yellow and white Wood Anemone.

This is the hexagram known as *Ch'ien* or The Creative. It is from the ancient Chinese classic of change, the *I Ching*. It shares its 'creation' sixness with Judaism, Christianity and Islam.

The beauty of the bees' comb (cells in which their embryos develop) is consistently and dramatically six or hexagonal. This is due to the absolutely minimal wastage of wax in the comb's construction.

Top. This symbol is the conventional representation of the natural directions of space. When magnetically charged it is of great value to those on board ships on a sea with no other guiding features apart from the night sky. Note the beautiful eightfold star in the centre.

Bottom. A Cosmos flower, clearly exhibiting a finite eightness.

Right. Here in Fez the author was shown a traditional Islamic flower garden made up of star octagons and cruciform beds, slightly raised to aid the gardener's back. Here cosmology and flower-growing are integral. See illustration 129 on p.102 in *The Art of The Islamic Garden* by Emma Clark.

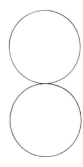

This is a classical twoness with an upper and lower circle. Both just touch at a single point.

The infinity symbol is like the figure 8 on its side. It has many depths of meaning should the reader be inquisitive.

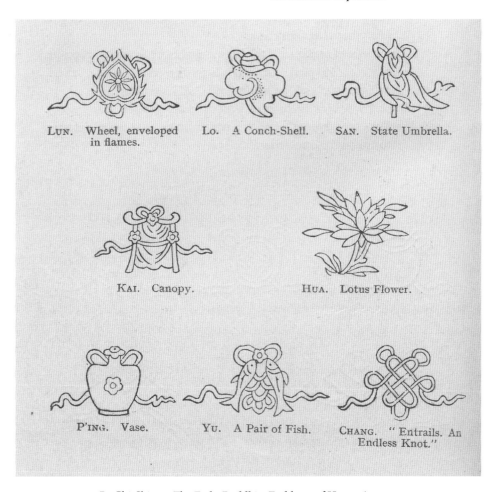

LUN. Wheel, enveloped in flames.

LO. A Conch-Shell.

SAN. State Umbrella.

KAI. Canopy.

HUA. Lotus Flower.

P'ING. Vase.

YU. A Pair of Fish.

CHANG. "Entrails. An Endless Knot."

Pa Chi-Shiang: The Eight Buddhist Emblems of Happy Augury.
Ancient China devised an array of eight things to satisfy each of the aspects of human life. This again is a beautiful seedbed of various meanings.

Cacti, most unusually in the plant kingdom, display beautiful examples of ninefold symmetry.

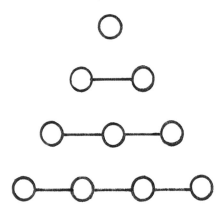

The ten circles in this drawing refer to the ten monads or 'ones' in triangular form. They represent the Pythagorean tetraktys or fourness (1+2+3+4=10) and thus the whole.

The Muslim genius in geometry gave rise to patterns never seen before in human pattern making. Here we see a tenfold central flower-like form surrounded by star pentagons. The pattern repeats if this rectangle is close packed with similar ones.

patterns. The Pythagoreans called ten the all-embracing, all-limiting 'mother' as it was an expression of the sacred tetraktys or fourness as shown above. In ancient Egypt we see the number 1000 (10x10x10) represented by the image of the Lotus flower. For Aristotle there were ten categories to cover all things and eventualities. The sacred *Rigveda* of ancient India is composed of ten books. Moses gave Ten Commandments which he received from God. Buddhism also has ten commandments, five for monks and five for lay people. The sacred and mainly esoteric *Kabbalah* of Judaism is structured on ten sephirots or archetypal emanations.

The whole of the created order can be taught on this kabbalistic 'tree' structure with its 'four worlds' or levels of being.[28] Annemarie Schimmel explains in her seminal book *The Mystery of Numbers* that when the Persian poet speaks of a Lily with ten tongues, he intends to say metaphorically that this flower has a great number of petal-tongues, by which it expresses its silent praise for God.[29] What is indicated here is ten as the number of perfection, not the literal number of petals. A Lily is most likely to have only six but four qualities in addition to these six!

At ten we have to bring our short summary of numerical 'nesses' to a close — although our illustrations show that we will find the petals of flowers exhibiting 13 and 21 in the golden sequence of Fibonacci, and many numbers between. Particularly the cosmic number of *governance* between sun and moon, or the twelve months of a year.

Petal numbers have ever been a fascination to people, not least the unrequited lover who plucks the daisy petals reciting 'loves me, loves me not, loves me, loves me not' until the end of the poor flower reveals the truth.

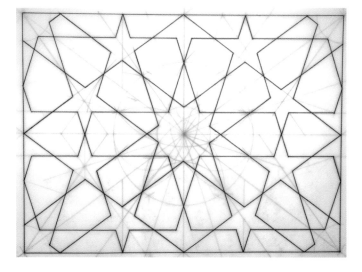

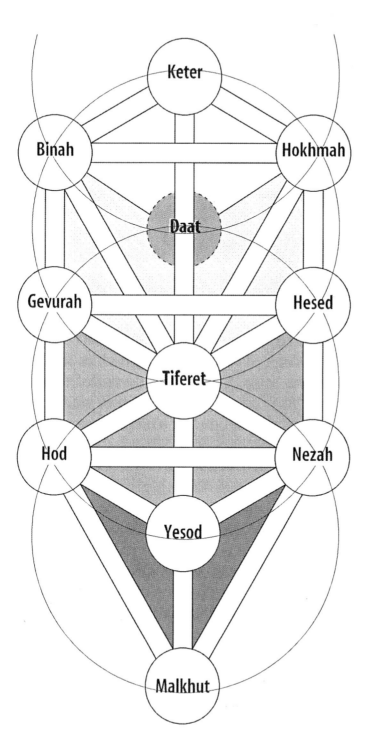

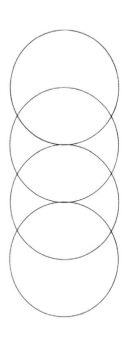

AZILUT - Emanation

BERIAH - Creation

YEZIRAH - Formation

ASIYYAH - Action

This remarkable philosopher's tool is called The Tree of Life. It is based on ten positions or sephirot. To learn more about this ancient wisdom see Ben Shimon Halevi's *A Kabbalistic Universe*.

The four worlds or oneness reflected four times.

Rhythmic articulation is beautifully displayed in the way the blooms of this flower emerge on the stalk. This pattern gives each the best opportunity to flower — singly and as a whole.

Rhythm and Periodicity
in Life and Flowers

. . . Look how the floor of heaven
Is thick inlaid with patens of bright gold:
There's not the smallest orb which thou behold'st
But in his motion like an angel sings,
Still quiring to the young-eyed cherubins;
Such harmony is in immortal souls,
But whilst this muddy vesture of decay
Doth grossly close it in, we cannot hear it.

Shakespeare, *The Merchant of Venice*, Act V scene 1

Numbers are the reverberation and pulsations of the One.

K. B. C.

Rhythmic periodicity is a characteristic of all phenomena as it is of all life. This is why *observations in time* for each plant and each part are so important.

We all know the great joys of what we call springtime particularly in the United Kingdom where the seasonal changes are so marked. But how many of us are aware that we also are being asked to wake up with nature? Farmers and flower growers certainly are aware of this natural law and rhythm.

Why do flowers so affect us? Is it only sentimentality or is it something more profound? After all we and they are part of the same living whole. Are there also seasons to consciousness? Why should we not consider the inner purposes of flowers to be our reminders, even teachers? The movement of living plants and animals is called their physiology, the effect on us can be called psychology or soul-changing.

E. Bunning said, 'Physiological research . . . is knowingly one-sided', in other words life is not just a complicated collaboration of substances and forces. The form and life of living creatures needs be seen as one unity. Sadly we have the dichotomy of morphology and physiology as quite separate studies. This is the result of an overemphasis on the principle of difference and the inevitable result of fractioning specialization. Bunning[30] and Kleinhoonte have been attempting to bridge this split

Honeysuckle. We 'see' leaves in a proportional display governed by the rate of growth in time. These leaves do not cover each other's access to light and so are mutually sustaining.

since the 1920s in their rhythmic research studies. They have demonstrated how rhythms of varying duration determine numerous metabolic processes from the movements of leaves and petals to cell division and propagation. Having posited the 'physiological clock' and observed the permeation by rhythm in so many aspects of all life processes, they eventually named this *the biological clock*.[31] The source and cause of this 'clock' still has to be detected *within* the bodies of the creatures studied. Is it thereby above and beyond the material plant? Or is it an outside influence that affects the total plant? It is a shame that this work on periodicity has fallen out of fashion. Do we automatically abandon what we cannot explain? Rhythm is a basic factor in the manifestation of life from heartbeat to peristalsis, respiration, cell division and so on. In short, life itself is positively rhythmic and cyclic.

An elegant and distinguished
tall flowering spiral.

Growth: the Floral Unfolding in Time

The tree is a winged spirit
Released from the bondage of seed
Pursuing its adventure of life
Across the unknown . . .

Rabindranath Tagore

The basic character of the living is to divide, to unite, to surrender itself
into the general, to persist in the specialized, to change, to particularize,
and as the alive might display itself in a thousand conditions, to step
forward and to disappear, to solidify and to melt, to conceal and to flow,
to expand and to contract.

Goethe, Theory of Colour 1:267

Although, as Goethe observes, life is multi-behavioural, it also follows distinct phases.

As we have already shown there are at least *seven* stages in the geometry and behaviour of a flowering plant.

1. The seed: as the *point* of departure.
2. The first shoot: as the *line* of development.
3. The stalk and the root 'shoot': as extended life *lines.*
4. The first leaves plus roots: as the unfolded *plane* of flatness.
5. The first bud of flowers + leaves + roots: as the *point* of solidity of the flower which will be housing the next generation.
6. The unfolding flower + leaves + shoots: as the fulfilling sphere of display and celebration of flowering.
7. The developed fruit and seeds + leaves + roots: as the consolidating sphere containing multiple seeds or points within the solidity of the seed pod.

This rhythmic and sinuous form exhibits the dance of life as it reaches towards the light and the changing position of the sun in the sky.

The fivefold Passion Flower bud releases its inner glories.

The above seven stages are dependant upon:

Light
Soil or darkness
Water, both above and below
Air: the reciprocal breaths of oxygen and carbon dioxide
Movement: fundamental to life and sap distribution

Inherent within, and intrinsic to, the above are ten further factors. These are both principles and physical factors.

a. Symmetry — balanced and poised activity. (A principle)
b. Geometry — structural and 'in principle'. (A principle)
c. Gravity resistance — impulse of life toward light. (A physical force)
d. Insect interaction — collaborative fertilization and nourishment. (A physical association of intimacy)
e. Fertility — necessary regenerative action. (A principle)
f. Colour — to attract attention, including the infra red and ultra violet ranges. (A physical factor of light)
g. Pollination — the subtle principle of fertilization. (Both physical and principle)
h. Fragrance and perfume production — attraction to collaborators. (Physical though subtle)
i. Animal and human welfare, medicine — natural health promotion. (Physical interrelatedness)
j. Animal and human nourishment — 'All flesh is grass'. Service and sacrifice. (Physical and principle, universal interdependence)

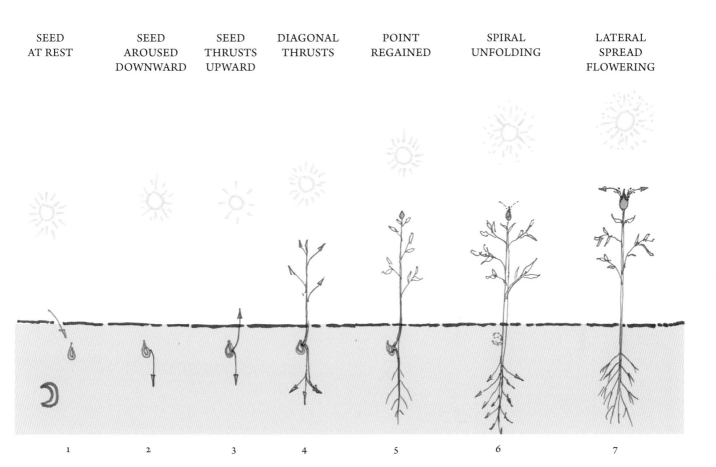

SEED
AT REST

SEED
AROUSED
DOWNWARD

SEED
THRUSTS
UPWARD

DIAGONAL
THRUSTS

POINT
REGAINED

SPIRAL
UNFOLDING

LATERAL
SPREAD
FLOWERING

1   2   3   4   5   6   7

1. Seed point of power and potential falls into or is planted in Mother Earth.
2. Seed first thrust to Mother Earth.
3. Second thrust up to the light and sun.
4. Third thrust diagonally outward both downward and upward.
5. Fourth thrust bud at apex repeats form of originating seed.
6. Fifth thrust bud begins its spiral unfolding pulled outward as well as upward by the Light.
7. Sixth thrust, the petal spirals outward to make a lateral plane to the sun. Then it usually follows the sun's path.

The common apple is a vital symbol in the Judaeo-Christian creation story.
Apples divide into five simple seed sections when cut. The geometrical
proportions are however extremely interesting as our analysis shows.

The Seed as a Point of Origin and Potentiality

If these green trees are heavy, their weight is in my hand.
Kathleen Raine, from *A Strange Evening*

Within the centre of the rose
Seed out of the silence grows . . .

Kathleen Raine, from *Out of Nothing*

Each life cycle, from the smallest to the great Oak, begins with a seed or a bulb, one that has normally lain dormant over the resting period, usually the winter. In some of the plants of our planetary home, however, the seeding cycle is virtually continuous due to the climate and seasonal conditions as well as an ability to lie dormant for more than centuries. Note the grain found in Egyptian pyramids and 3,000-year-old Lotus seeds in Japan. All these proved fertile and were propagated.

A seed is the normal point of departure of the life cycle of any plant and it can be considered the intense compression of all the plant's potential — a dynamic mystery of intense contraction.

We have become remarkably ignorant (in modern agribusiness terms) of the reliance of seed germination on the moon cycle which is fundamental to all traditional farmers and such as the Hopi Native Americans who live in the most botanically hostile region of the Mesas of Arizona. Here they rely on their traditional knowledge of seed orientation and the moon phases.

The author was fortunate enough to be tutored in some of the closely-guarded secrets of how the Hopi people (with the longest-known human history of *non-warfare*) planted according to the moon phase. He was also taught how they store their seeds in special sanctuaries (*kivas*) below ground with their radicals aligned to the geomagnetic field of the planet. Such esoteric knowledge is vital to the survival material of these people who have kept their land because no others could get the corn to grow in such a 'hostile' environment. Their secret, it seems, lay in the treatment of seeds. The orientation of dormant seeds and their increased fertility

So much is contained in this seemingly simple cross-section of the common apple. The seeds are arranged in a star pentacle. The profound proportions are shown on the opposite page.

FRENCH BEAN SEED GROWTH

This illustration represents the coming into being of a plant form from seed to the fruit shoots and stalks. The seed, egg-like, is usually dormant for the winter period. With the warmth and moisture of the spring it is likely to germinate. The point-seed-origin signals this by putting out its first shoot which is linear. As soon as the first linear shoots move heavenward (to the light) then the second linear movement proceeds to move toward or into Mother Earth or the darkness of mineral nature. Thus the two lines become one with the seed point in the centre. At this point more shoots push out toward the life-giving light and prepare for the miracle of planar leaf form which is the theatre for photosynthesis. In the meantime the earth shoots expand and multiply to ensure the necessary water supply. (Liquidity or water is the essence of all life whether it be sap or blood or any of the other life-supporting liquids. As it says in *The Holy Koran* 'We brought through water every living thing'.)

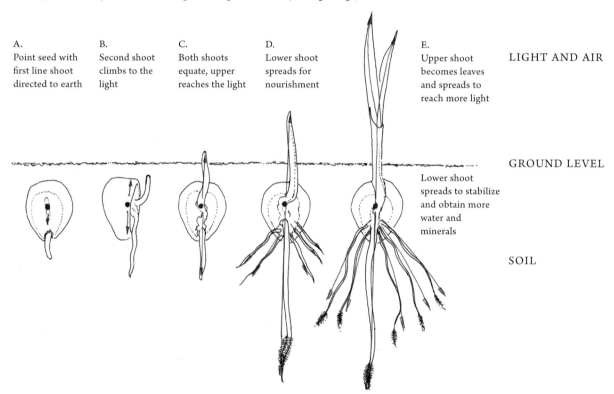

A.
Point seed with first line shoot directed to earth

B.
Second shoot climbs to the light

C.
Both shoots equate, upper reaches the light

D.
Lower shoot spreads for nourishment

E.
Upper shoot becomes leaves and spreads to reach more light

LIGHT AND AIR

GROUND LEVEL

Lower shoot spreads to stabilize and obtain more water and minerals

SOIL

The process of starting from a point of origin as the seed body, followed by the shoot moving down to Mother Earth and up to the nourishing daylight. This shoot eventually becomes the stalk on which the leaves unfold — until it eventually produces the flower which will become the fruit and thereby the new seeds.

a. The point sphere or the seed first sprouts into Mother Earth darkness. *Life begins in darkness.*

b. Next, a linear shoot is attracted to find the light and sustainer of life. *Water and minerals are needed from below.*

c. Now the linear shoots are more balanced. The light is reached. *Life resides between heaven and earth.*

d. Greater strength is required from Mother Earth, both nourishment and stability.

e. Life continues to take possession of space. The root's world remains linear, the light-seeking becomes planar. The roots below become more numerous and specialized to take up their nourishment, minerals and water — but they essentially stabilize the wind-blown upper leaves.

was scientifically demonstrated by the Russian Academy of Science in the 1940s. It was called 'botanical geomagnetic responses'. The correctly oriented seeds were many times stronger and more fertile than randomly oriented seed. This the Hopi had known for centuries by intelligent practice.

Watching the seeds grow, being aware of the conditions of existence they move through, is a knowledge well known to all traditionally wise agriculturists and farmers. This wisdom is continually under threat from the commercial forces that 'engineer' seeds and their fertility for short-term financial gain and have caused so much distress and misery to traditional farmers. HRH The Prince of Wales has continually pointed out this tragedy.

The flat or planar leaves perform the miracle of photosynthesis, the transformation of light into the sugars of nutrient for all animal life as well as for the plants themselves. The geometry of leaves is an equally wondrous study. Leaves are presented to the sky as receivers of light. Their variety is astonishing, even if we in our relative ignorance think that all they do is photosynthesize. Who can explain the shape of leaves?

Above the leaves which nourish life comes the site of 'flowering'. This includes the principles of pollination and fertilization from which comes the resultant growth of the fruit, and finally the seeds from this fruit. Pollination is virtually microscopic and another engaging mystery in the floral world.

To scrutinize the *form* of seeds is particularly important geometrically. In an archetypal sense we can suggest that the sphere is the shape of perfection. As the sphere becomes a moving form it changes into an egg-like shape with more than a single centre. Seed forms usually include both sphere and lineation and vary as much as flowers' forms (from the spherical pea to the elongated bean). Seeds unfold in a geometrical series taking possession of space until they recover their spherical form in fruit like the apple, peach or plum.

Like all else, for a seed to achieve this pattern of growth it needs to enter the 'conditions of existence'. Thus the conditions, whether or not simultaneous, which we have already proposed are: spatial, temporal, formal, numerical and substantial. This last is the principle of phenomenality itself. However it must be borne in mind that these conditions are in themselves pure principles which yet 'inform' the domain of the changing world of phenomena. As growth acts in the *changing* world there is no possible finality that can be posited about its patterns of behaviour; hence it has been placed in the domain of the 'seeming', 'likely' or 'illusory' by the sages of past time. Buddhists particularly

Orange

Plum

Peach

Avocado

The nature and comparison of seed forms. An extensive study in itself but there is clearly a common characteristic in all the varieties. The *vesica* character is always evident, though one end may be more rounded than the other.

The *Kiva* is the sacred space of certain North American Indians where during the winter season corn seeds are aligned to the earth's magnetic poles. This way of reinforcing the strength of the seeds was confirmed by the Russian Academy of Science during the Second World War. They called it 'magnetotropism'.

advocate complete detachment from this 'seeming' domain so as to find the unchanging heart of the matter which they see as the ineffable void within. Thus *nothing*, like zero in mathematics which is fundamental to all numerical calculation, must be the next issue to be *recognized*, *defined* and *understood* by the responsible contemporary scientific community.

To remind us of the internal meditative wisdom of Buddhism we have what the Zen Master Takuan Soho said in the sixteenth century. 'It is the same and at the same time is not the same, it is different and is not different'. Master Takuan called this 'eternal immortality' and said, 'a flower blooms eternally and leaves fall eternally'.

From Bud to Flowering

'Now that', she said, 'which seeks to subsist [be] and endure, desires to be one; for if this unity is destroyed not even continued existence will be left to anything'.

Wisdom herself talking to Boethius (in *Consolations of Philosophy* III 105)

From the very first indications on the twigs or between the leaves on a stalk we can detect that a flower is due to arrive. For the wise Goethe this represents a significant transformation of leaf into a new complex and colourful form, the flower. He saw the flower petals as new coloured and transformed leaves.

The flower bud as it arises soon indicates the symmetry that the petals will display. This is a fundamental mystery as many flowers do not adhere to only one symmetry. In short a particular plant can become four-petalled, five-petalled, six-petalled and even seven-petalled from the same parent body or even the same stalk! (The Primrose family is one such example.) Does this not challenge the sole control of the genes? If not, what could be the reason or function of multiple petal symmetry?

Overleaf we have taken a series of images of the opening of the *Balloon* flower. The first 'closed' bud is fivefold and will eventually flower into a five-petalled blossom. The remainder are four-petalled in different states of unfolding. The three-dimensional shapes described by both the body of the unopened bud and the 'path' that the petal tips describe are all *of the utmost beauty*. Is the sunlight pulling the flower into shape or is the inner life force of the flower 'pushing' it into shape, or naturally and paradoxically both? Neither, however, appears to determine the consistency of each flower's symmetry. The Balloon flower appears to exhibit individual choice — or are there other factors beyond our recognition so far?

This Bougainvillaea displays the Goethean transformation of leaves to flowers. In fact the flowers themselves are quite tiny and fivefold but the leaves become coloured around them.

FOURFOLD SYMMETRY which is the predominant symmetry of the Balloon flower.

FIVEFOLD SYMMETRY

A

B

C

TWO SYMMETRIES OF THE BALLOON FLOWER

Meditation on the life-cycle of the Balloon flower is strongly recommended. This lovely smallish flower is mysteriously choosy about which symmetry it wishes to display for its petals. But choose it does.

On the fourfold side:

1. It produces a fully blown balloon.
2. Its first opening.
3. The remarkable elegance of its profile is displayed.
4. Another view of the next stage where the petals begin to curl the way the opening is moving.
5. The fourfold (square) opening is beautifully poised.
6. A full-face image of the fully opened flower: note the central pistil follows the symmetry of the petals.

The fivefold side follows in a similar manner (see ABC).

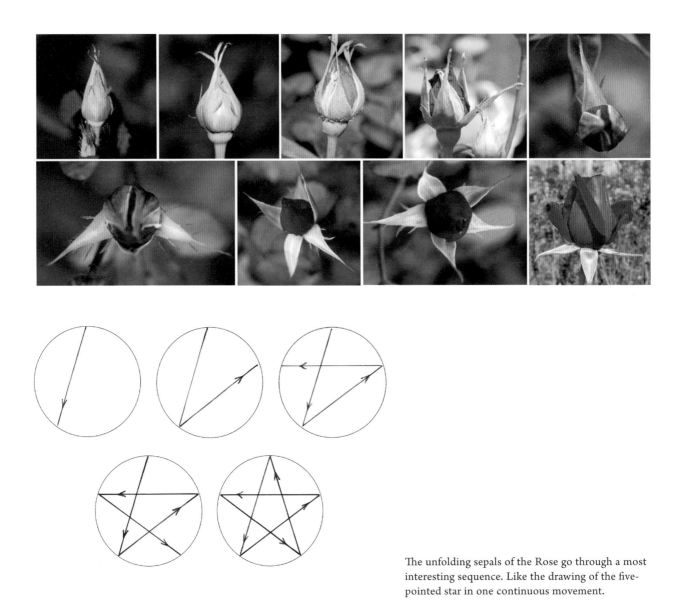

The unfolding sepals of the Rose go through a most interesting sequence. Like the drawing of the five-pointed star in one continuous movement.

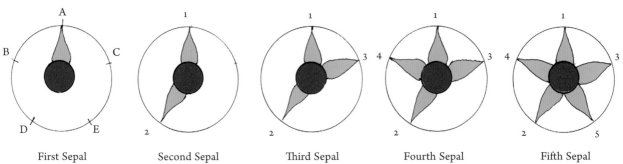

First Sepal   Second Sepal   Third Sepal   Fourth Sepal   Fifth Sepal

Engaging With or
Talking About Any Given Flower

To the one whose soul is musical, who sees everything as spatial, figurative and shaped, forms appear through unconscious vibrations. Whoever sees musical sounds, movements etc. within himself, his soul is spatial because the variety of sounds and movements only arises through figuration.

Novalis, from *Fragments* I

Taking a flowering plant as a whole it can be embraced by a single title 'Daffodil', 'Snowdrop', 'Pansy', and so on; but it is important to remember that spoken language and naming is strictly a human impulse, albeit of the greatest importance when needing a common or universal language to communicate with. Yet the flower never 'names' itself, it simply displays itself in its own time, rhythm and place.

When Adam was instructed to name all the creatures, this symbolized the importance of finding words to describe things or creatures which are not immediately present. It is the prime method of abstract thought and communication and that which qualifies us as *human* animals. Names are of the intelligible domain and inevitably abstract. They become cultural conventions for each civilization.

A flower is primarily a three-dimensional form which, when studied, goes through a magnificent series of three-dimensional geometric movements. Some plants contract into a form that revitalizes the next year and some die, which is the withdrawal of life. The seed (or point) is the ultimate contraction into intense potential, resting or dormant until the next season of growth. So the flowering plant needs be appreciated as a whole at any stage in this lifespan.

Thus we can see the analogy between the plant taking possession of space and the *geometric dimensions* arising from the 'movement of the point'. Life is synonymous with movement. Here the metaphor is the point moving to 'create' the dimensions of space or the seed unfolding into the life of the plants and taking possession of space.

Here we have characterized the point as a point/sphere: this is so as to appreciate that if it has come into being it will from the start contain all the dimensions. (See *Order in Space* by Keith Critchlow).[32]

Water Violet. A member of the Primrose family.

The unfurling to the light from a beautifully integral form. Like a seed in shape, the green bud divides to open to the life-giving sunlight. The elegant laws of spiral unfolding are in attendance.

We can also draw an analogy with the point/sphere of the emergence of a potential flower by calling the flower *bud* a further 'point of departure'. This is a departure in the sense that prior to the emerging flower bud we could see only leaves in their planar unfolding.

Within this growing flower bud are the enfolded petals of the emerging flower. These are neatly packed in different ways — most often in a spiral manner of some kind. In the case of the Daisy petals for instance they are also 'rolled' into linear features prior to unfurling into flat petals. These intensely wrapped petals or sets of petals proceed in a linear fashion to emerge out of the 'bursting' bud which is composed of an outer shell of sepals. In the Rose for instance, these sepals are five in number and burst open in a sequence that naturally makes up a rhythmic five-pointed star, in the most balanced way.

Another 'pattern of life' with a geometric and arithmetic character is the spiral. This is a factor often in attendance when sepals or petals unfurl. These botanical patterns most often demonstrate a predominance of what has come to be called the Fibonacci or 'golden' sequence. This sequence is so beautiful in its simplicity and profound in its proportions. It is both hidden and revealed simultaneously.

This 'golden ratio' is immediately recognizable as the mysterious geometric proportion between all the divisions in the five-pointed star as we have shown. Fiveness, the golden sequence and life in nature are integral and inseparable though not exclusive.

Thus we soon see that living creatures, particularly those with easily recognizable fivefold symmetry, are demonstrably comfortable in 'using' or expressing this golden sequence. For we, as humans who analyse such behaviour patterns, realize that the sense or recognition of beauty is inseparable from these mathematical sequences and proportions. As we have already said, in Plato's *Timaeus* we learn that it is proportion that coheres the whole universe and that this was called the 'world soul'. Therefore we should not be surprised to find the use of what is called the 'golden proportion' in all civilizations in their sacred art forms and architecture. In particular that which inevitably reflects their cosmology or understanding of universal law.

To be 'in tune' with timeless truths is the objective of the quality we called sacred. The author's provisional definition of the word 'sacred' is: 'that which is essential to human life and thereby all life'. Temples or sanctuaries, places of prayer and ritual, have been consistently an integral part of all human groups throughout history — as the invisible world of spirit has always been considered integral to, and inseparable from, the material world — in fact its parental cause. 'Now to discover the Maker and Father of this Universe were a task indeed: and having discovered Him to declare Him unto all men were a thing impossible', as Plato's Timaeus says.[33]

It is the 'evidence' staring us in the face, in the unfolding of the flowers, that convinces the author of the profundity of the blossoming world and its integrality with human life. This is quite above and beyond a flower's material value as food or medicine or even simple pleasure (which is also another dimension of our inseparability). Flowers nourish us at all levels if we let them and are awake to them.

That which we cannot explain we tend to call the 'aboveness' and 'beyondness' which is, after all, represented by their beauty — as expressed by the geometry, symmetry, numerical sequencing, colour, form and perfumes of each flowering plant. A beauty that points to the 'Beautiful Itself' as Plotinus characterizes it.[34] We find Socrates in an unusually affirmative mood in the *Republic* when explaining to Glaucon that geometry is 'of the knowledge of things which exist for ever, rather than things which come into existence at some time and subsequently pass away'. 'Therefore, Glaucon', continues Socrates, 'geometry can attract the mind towards truth. It can produce philosophical truth . . . it can reverse the misguided downward tendency we currently have'. Socrates was alluding to the origin of the word *geo* meaning earth, looking groundward rather than heaven-ward. The study of geometry, Socrates insists, 'makes absolutely all the difference in the world'.[35]

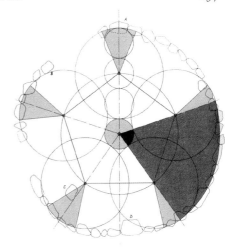

That our ancestors were not only knowledgeable but practitioners of literal geo-metry is evidenced in the Stone Temples in the United Kingdom. The stone spheres of this period confirm their sophisticated knowledge. See *Time Stands Still* by Keith Critchlow. This is a geometrical analysis of the Welsh stone circle Moel Ty Uchaf by Professor Alexander Thom.

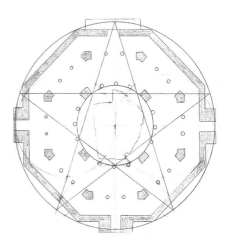

That a particular aspect of the proportions within a pentacle or five-pointed star was used throughout history by the masonry guilds is evidenced in the proportions that are inherent in the Dome of the Rock in Jerusalem. This analysis is from Dr Emily Pott's doctoral thesis, presented to the University of Wales from the Prince's School of Traditional Arts, London.

A GRAPH OF THE 'GOLDEN' OR FIBONACCI SERIES

It was not a coincidence that Plato did not mention this golden ratio and its relation to the dodecahedron (cosmic regular figure) in his cosmogony, *The Timaeus*. It is so intimately connected with the journey of the soul — which Plato did concede was composed of proportion. There is still much to learn here.

As the soul grasps the spiritual significance of this golden number it approaches nearer and nearer to the ineffability of the source itself.

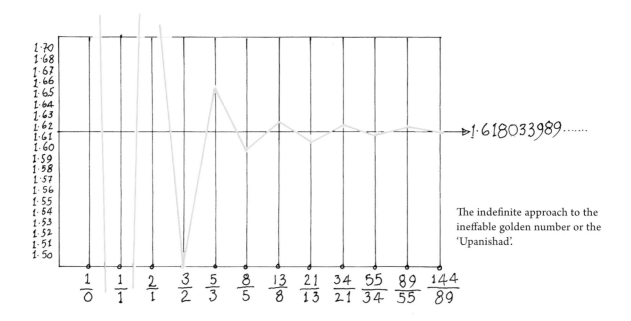

The indefinite approach to the ineffable golden number or the 'Upanishad'.

The natural oscillation of the progresive golden series of whole number proportions. Each number is the sum of its previous two digits.

The numbers in this series conform to the universal human sense of beauty in sound:

Pentatonic scale: (5)
Diatonic scale (8)
Chromatic scale (13)
Shruti scale from Vedic Indian music (21)
2:1 = The octave
3:2 = The musical fifth and so on

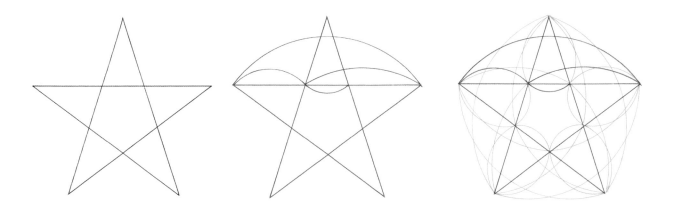

In the star pentagon (left) we can draw the proportions between the divisions of
the arm length as shown by the curved lines (centre). All the proportional curves
can be systematically filled in (right). The result is a fivefold flower.

As the DNA molecule, when cut across, exhibits
a tenfold symmetry we can expect to find golden
proportions within it. This is shown on page 199.

Repeating patterns in fivefold symmetry were virtually
unknown until the Islamic geometric genius first explored
them. Here a beautifully simple pattern is composed of regular
pentagons and what can be called the vessel shape. The
angles of this classical shape are composed of the two golden
rhombs more recently associated with Professor Penrose. By
connecting the pentagons' centres with a dashed line, special
close-fitting hexagons emerge.

Two principles are shown simultaneously. (1) Packing fruit together for storage.
(2) The packing is hexagonal with a central seventh.

The Singularity of Fruit and Seeds

Sensitivity to see beauty, to enjoy the loveliness of an evening, to look at a tree and be intimately in communion with nature.

J. Krishnamurti, from *On Pleasure*

One gives birth to Many and Many strives to recover Oneness again; this is the rhythm of life.

K. B. C.

As the flower is the crowning celebration of the plant's primary achievement, there follows the necessity of having to facilitate the coming generation of the same plant. This is the parenthood of the flowering plant. Achievement, in this case, is the accomplishment of a fruit filled with the seeds for producing the next generation.

Naturally we cannot overstate the importance of the fruit or seedpod which is an essential aspect of the miraculous continuity of life. Fruit has more than a single function as we are all aware. As the flower can be considered nourishment to the soul in its beauty, perfume and colours, so the fruit is nourishment to the other members of the ecological community such as the insects, the birds, the animals and the human family. We only need to cite the apple — so fundamental to Christian symbolism. The desirability of fruit increases the distribution of the seeds in another mutual collaboration.

Humankind has cultivated key fruit-and-seed-yielding members of the plant world, from the cereals of corn, rice, wheat, barley and oats to

Here we show an apricot seed between its divided halves. The 'flesh' that can feed so many other creatures is part of the strategy of being taken away to eat, with the 'stone' discarded sometimes many miles from the original tree.

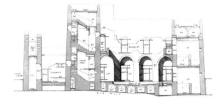

This section through an ancient Egyptian grain store shows how sophisticated the Egyptians became, both architecturally and as storers of grain.

Civilization becomes possible when humanity finds a strategy to store grain to feed itself through the non-growing season. This is from an ancient Indian civilization.

As the longest continual human civilization the ancient Egyptians agricultural wealth related directly to the flooding of the Nile and the subsequent grain growing in the nutritious silt that came with each season.

such vegetables as the pea. It can quite literally be pronounced that *civilization is not possible without a surplus of both grain and fruits*. So human fulfilment and the realization of becoming fully human is utterly dependent on the reliable and sustained food source. This is the contribution of the plant world. How could ancient China have achieved such a magnificent civilization without rice or bamboo? Civilization has to be seen as a collaborative partnership between the plant and human kingdoms. All great civilizations have acknowledged the sacredness of natural foods and demonstrated immense ingenuity devising magnificent storage vessels, whether sophisticated or organically utilitarian.

It hardly needs to be repeated that we are currently at a dangerous point in human development because we are forgetting the most fundamental facts of healthy food production: the health of the soils, the abundance of the insects and the health of the vegetative world, to all of which we owe our lives. Mechanization and chemicalization of food production are self-defeating and destructive to a large range of living creatures, ourselves included. Thankfully some of us are being awakened to this currently. Not least by HRH The Prince of Wales[36] who has championed the Soil Association, the green activists and others in the UK and Europe. Also we are thankful for the life work and insights of the Americans Wes Jackson and Wendell Berry. The latter is both poet and farmer.

Here is a magnificent African grain storage unit, not unlike a traditional beehive.
Note it is on legs to discourage the rats from being fed as well.

Pollen, beyond our vision as single particles, truly marvellous in its geometric sophistication.

We *Are* Nature
As Well As Supra-Nature

You are the mighty way
in which every thing that is in the heavens,
on the earth, and under the earth,
is penetrated with connectedness,
is penetrated with relatedness.

Hildegard von Bingen

With a fiery spirit
I transform it
into a body
to serve
all the world.

Hildegard von Bingen

The symbiosis of the natural world and humankind has been the basis of successful sustained human life since recorded and pre-recorded time. We *are* nature, we are the 'conscious' living cells within our mother planet — to divorce ourselves from nature is to *divorce ourselves from ourselves,* in short a suicidal act.

As there are levels of awareness so there can be the metaphor of food or nourishment for each level.

Bodily food for the body.

Social food as compassion for societies.

Cultural food as enshrined in the arts of each culture.

And spiritual nourishment nurtured by intuition, inspiration
    and faith.

Each can blossom and bear fruit when seen as an inseparable part of the whole. All these factors can be identified in the comprehensive history of life and ourselves as the human ingredient of life. The systematic separation of these same human necessities naturally becomes another form of destructive atomization or fractionality. Analysis 'dissects' symbolism and synthesis 'throws together'. Life harnesses a harmonic balanced rhythm.

Here the Moon Daisy (Marguerite) displays the great geometrical secret of the golden spirals.
It is only persistent scrutiny that lies between anyone and this profound open secret.

The Unfolding of a Daisy, a Symbol of the Flowering Process Itself

From forms that transcend matter, have come forms which are found in matter and effectively constitute bodies.

Boethius, from *De Trinitate*

Here we have chosen stages in the unfolding of the common Daisy as a metaphor for all flowering.

To ignore the question of what flowers do for us would be a lamentable failure. We are all aware of how powerful the value of flowers can be and always have been in the celebration of birth, rites of passages, marriage, maturity and the passing on to another existence in the beliefs of the majority of human kind.

To describe flowers purely 'rationally' is a further example of the trivialization of natural beauty let alone the display of the unfolding which is the celebration of all living things.

Here the process of emergence of the Moon Daisy has a particular message in terms of form. From point sphere-bud, to the natural 'break out' into line petals which become flat petals symmetrically displayed until the central disc of spiral formation comes into view. The vortex from below spirals upward to eventually become a hemisphere, the petals then turn under to repeat a larger version of a full sphere in principle.

At any given moment in a Daisy bed we can see all stages of growth simultaneously.

MOON DAISY SEQUENCE OF FLOWERING

Because we are so intent on seeing a flower at its best, we rarely pause to follow the bloom from its first little bud through the series of movements it forms up to the final opening and then its withdrawal. Here we have recorded the Moon Daisy's 'dance' into flowering.

a. The first minute indication is a little spherical bud between the leaves.
b. The bud grows radially yet prompted by the number of binding sepals.
c. The bud at its greatest is like a slightly flattened sphere.
d. If we look on top of this tightly bound sphere we see it has a transparent 'window' to the petals within.
e. As the light seems to prompt the petals, they bulge outward spreading the sepals.
f. The bulging petals are seen to be wrapped invertedly in a spiral form.
g. The spiral indicates the way the petals are going to emerge one by one.
h. Like fingers unfolding, the small linear petals unfold in a spiral motion.
i. They reach upward still following their spiral form and create the rhythm of growth.
j. Looking down into the unfolding petals we see they leave a hole within which we can see the yellow centre.
k. As the petals reach toward the light each unfurls from a line to a flat plane.
l. As the light catches their translucent whiteness, these flat petals (still spiralling) create an overall cylinder.
m. The larger the petals get the more we see how they are beginning to flatten out. Now we have a cone.
n. The cone spreads, revealing more and more the yellow centre continuing its vertical movement.
o. The flower has now achieved its ultimate visual spread and has become a flattened disc attracting the most attention.
p. During this period of flatness, the yellow centre, which contains many smaller flowers, begins to fill out into a hemisphere.
q. The 'dance' of the petals seems to be celebrating the flowering.
r. The next stage in the life of a Moon Daisy is its petal contraction.
s. This leaves the maximum exposure of the seeds in the flower head and the completion of the special form of the whole.
t. At any given time of the Moon Daisy's flowering season one can see all the stages of unfolding simultaneously in the neighbouring blooms.

St John's Wort. Each specimen flower harnesses fivefold symmetry in its own way.

Nature seems to take advantage of the simple mathematical representations of the symmetry laws. When one pauses to consider the elegance and beautiful perfection of the mathematical reasoning involved and contrasts it with the complex and far-reaching physical consequences, a deep sense of respect of the power of the symmetry laws never fails to develop.

C. N. Yang, Nobel Prize Lecture

Like a beautiful flower, full of colour and full of scent,
are the pure and fruitful words of him
who acts accordingly.

The Dhammapada, translated by Irving Babbitt

The next principle that proceeds in the unfolding flower is a greater or lesser degree of *spiralling*. This turning into being, as it might be called, is very marked in some flowers, for instance the St John's Wort and the Periwinkle, and far less in others such as the Water Lily and the Passion Flower (*Passiflora incarnata*). Roses show a special preponderance to unfold in a spiral fashion which, for many, is a major feature of their supreme beauty.

We find a large range of symmetries throughout the flowering domain when the petals are fully extended and the flower has its full face open.

FOUR SPIRAL LINES FROM
CENTRE TO PERIPHERY BASED
ON AN EIGHT DIVISION OF
THE CIRCLE

Such spiral forms can be observed in two dimensions in a flat flower's face and also in the three dimensions of petals still awaiting their unfolding.

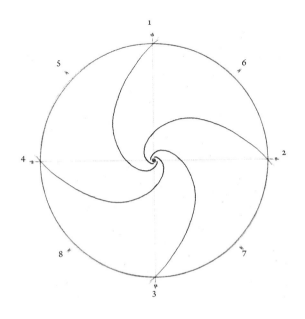

The Spirals of Unfolding

Grammar of five-fold rose and six-fold lily,
Spiral of leaves on a bough, helix of shells,
Rotation of twining plants on axes of darkness and light,

Kathleen Raine, from *Word Made Flesh*

Illustrations of spirals are inestimably more powerful than any attempt to be precise in words. One would say that life 'turns' into being. Here we have chosen to let the symmetry speak for itself.

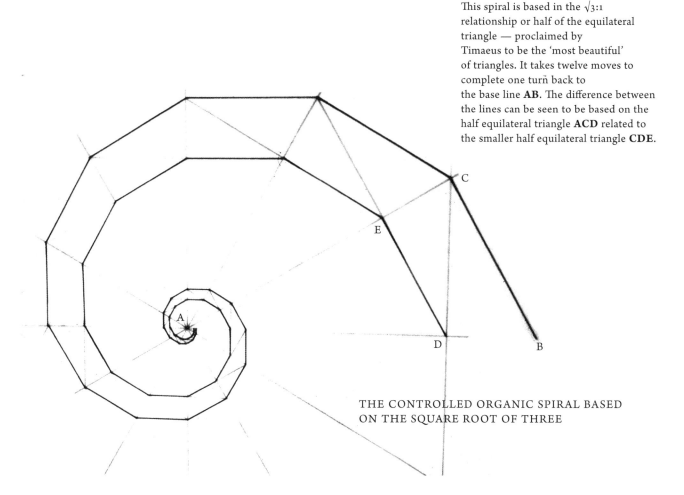

This spiral is based in the √3:1 relationship or half of the equilateral triangle — proclaimed by Timaeus to be the 'most beautiful' of triangles. It takes twelve moves to complete one turn back to the base line **AB**. The difference between the lines can be seen to be based on the half equilateral triangle **ACD** related to the smaller half equilateral triangle **CDE**.

THE CONTROLLED ORGANIC SPIRAL BASED ON THE SQUARE ROOT OF THREE

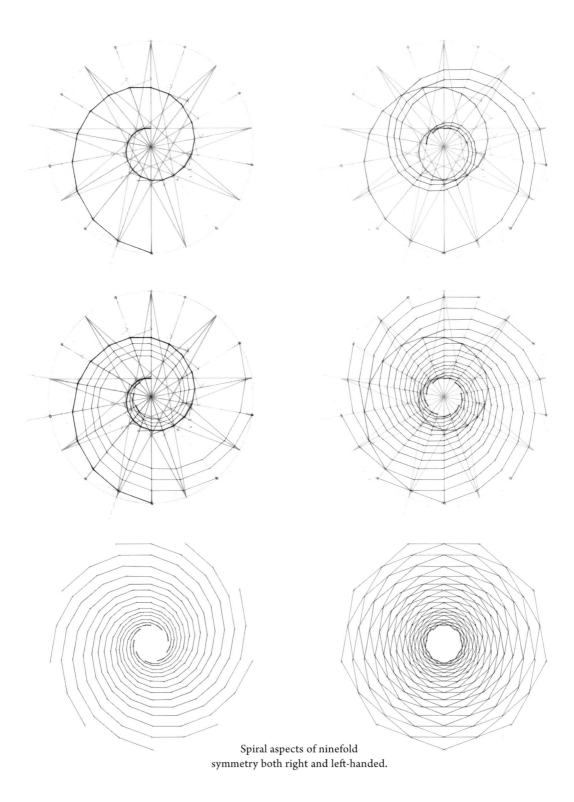

Spiral aspects of ninefold
symmetry both right and left-handed.

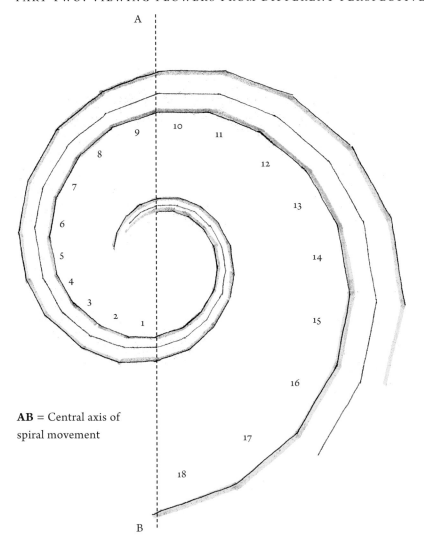

AB = Central axis of
spiral movement

AN ORGANIC GEOMETRIC SPIRAL BASED ON
THE ENNEAGON OR NINENESS

A spiral based in the nine-pointed star can be shown to take nine stages to get
back to the same axis and eighteen stages to make a full turn. Each archetype
expresses both its basic number and its own intrinsic form of spiral.

THE SPIRAL BASED
ON A NINENESS

Based in this instance on the ninefold
division of a circle, the spiral is as much
a mathematically factual phenomenon as
it is a metaphor and symbol of paths of
energy or botanical consciousness.

Proceeding from centre outward is
evolution.

Proceeding from periphery to centre is
involution.

The first develops or dissipates; the second
concentrates or integrates.

The spiral can move either left-handedly
or right-handedly. Here it is moving
inward left-handedly. It is also clear that
there are nine starting points for possible
identical spirals and these can cross over
each other with a symmetrical set of nine
right-handed spirals.

The spiral figures in most organic and
natural unfoldings in the vegetable
and animal worlds. It is a metaphor of
controlled growth, each new stage exactly
in proportion to the previous one.

Whether one approves of the cyclotron or not, at least we can admire the way the particle collisions demonstrate the spirals of decay at sub-atomic level.

Smoke in air moves in beautiful spirals.

Each part of the cauliflower exhibits its whole spiral nature rather like a chaos diagram — spirals within spirals.

In the spiral unfolding of its petals this beautiful flower demonstrates a principle similar to that of the spirals of growth shown above.

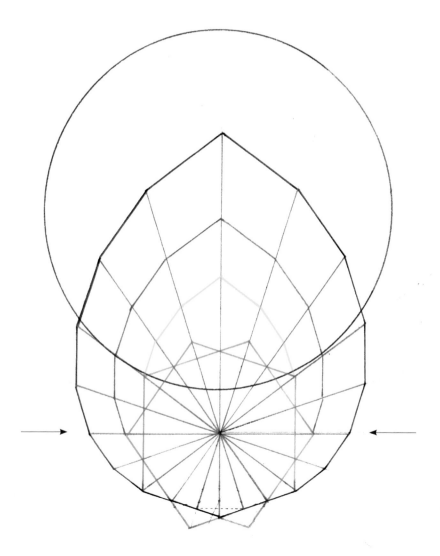

The decagonal star as the controlling archetype of a set of growing petal leaf or petal forms.

Each side of this leaf-form is based on a tenfold sequence from vertical axis back to vertical axis.

Three different sizes of this leaf-form are shown, all generated from points of the bottom central star decagon or ten-pointed star.

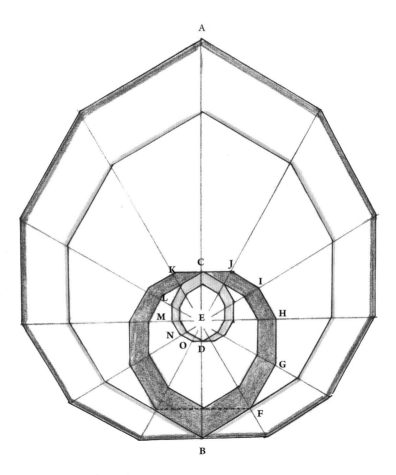

THE COLLABORATIVE EFFECT OF TWO √3 SPIRALS MOVING TO THE LEFT
AND RIGHT. A METAPHOR FOR PETALS AND LEAF FORMS IS EVIDENT.

The spirals can be seen to begin at **A**, cross the axis again at **BCD** and **E** each passing
through the central vertical axis. Proportional leaf or petal forms are hinted at in this drawing.
AB represents the dimensions of the first, **BC** the proportions of the second, **CD** the proportions
of the third and **DE** the proportion of the fourth.
The outer perimeter of the largest leaf form **AB** can be seen to be part of a
continuing spiral. This continues from **B** to **F G H I J K L C M N O D** and so on.

Thus it is only the polarity of a single progression that creates the effects of
potential leaf-forms.

These leaves will be found by any visitor to the J. Krishnamurti
Study Centre's sun-room (near Winchester in Hampshire).

Here the single-petalled Arum demonstrates
its spiral genesis.

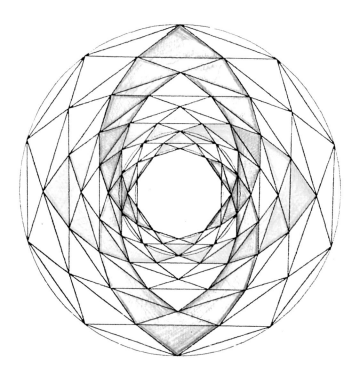

TWO KINDS OF 'LEAF' FORMS RELATED AT 180° TO EACH OTHER WITHIN THE DECAGON

This drawing is suggestive of the development of a 90° phyllotaxis, or a pair of leaves spreading alternately on their stalk at 180° at each pairing.

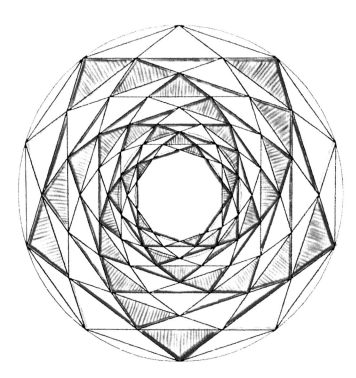

Each of the five coloured spirals, in this case red triangles, diminishes in a sequence of eight. If the drawing was complete they would reach the axis after ten stages

The unfolding of life in the plant world always expresses itself in spiral form.

This can be taken as the collaboration of space and time harnessed by the plant world for its mutual benefit.

The most significant spiral form could be claimed for the golden spiral which emerges from the five-pointed star.

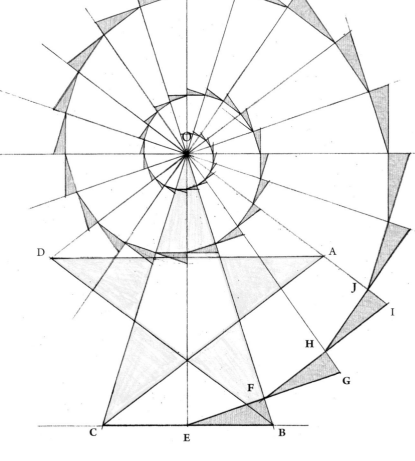

Here we have taken a simple example of a natural fir cone and coloured each seed piece either yellow in sequence or blue in sequence. Each sequence, left-handed or right-handed, crosses the other series to complete the cone. Here we have the Fibonacci numbers, 5 yellow and 8 blue, meeting or collaborating.

This drawing demonstrates the time factor in the delineation of the golden spiral, growing or diminishing from the base of the star.

The foundation of the golden mean spiral can be found in the star pentagon or pentacle. In our illustration this is represented by **O-A-B-C-D**. The point **O** remains as the focus of the spiral in time.

To show how this naturally moving spiral can be constructed we begin by dropping a perpendicular line from **O** to the base line between **C** and **B** at **E**. From the central bottom **E** we find the point **F** which lies in line **OB** at right angles to it. Next we draw the line from point **E** through point **F** and extend it to point **G**, so that **EF** equals **FG**. This results in the triangle **OEG** being exactly similar to triangle **OBC** but proportionally smaller. **OF** is the bisector of this triangle **OEG** so it becomes the generating point to extend to the line **OG** at **H**. This point **H** lies at right angles to the line **GO**. This line **FH** is extended to point **I** so that **IH=HF**. Thus we have completed the next proportionally diminishing similar triangle **OFH** and so on until a significant amount of the golden spiral is constructed. Here we have constructed fifty stages in the spiral.

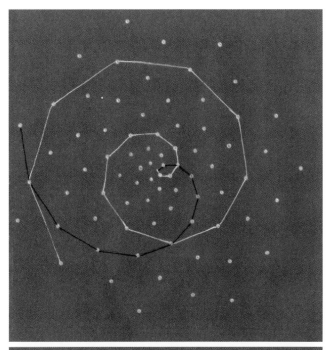

A is a left-handed spiral based on a displacement of 99.5° between points which leads to a whole number series approaching √3.

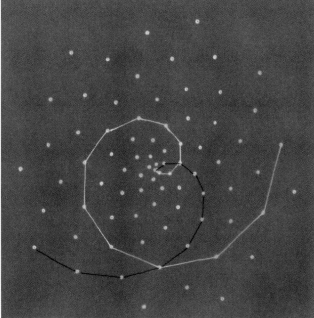

A right-handed spiral based on a displacement of 137.5° between points of juncture. This leads to a whole number sequence related to the golden section or $\frac{\sqrt{5}+1}{2}$

As this flower opens, it lifts its head to look out horizontally.

The Orientation of the Face of Flowers

Though we travel the world over to find the beautiful, we must carry it with us, or we find it not.

Ralph Waldo Emerson in his essay *Art*

As we have already said flowers face three possible ways: (a) to the sun meaning upwardly, (b) horizontally towards each other and (c) some flowers drop their heads and face mother earth.

Naturally there is beauty in the whole of a flowering plant — even if it is not our favourite. The beauty in the face of a flower is where we most expect to find it. This raises the question of the significance of flowers to ourselves. What empathy do we have for the plant world and even *why* do we have empathy for flowering plants at all?

Its range of qualities immediately lifts the flower above being merely an instrument of reproduction. The word 'functionality' has buried too many deeper clues, not least the key qualities of order, proportion and beauty.

The word 'face' is an interesting one that goes deep into symbolic as well as technical meaning. We are not likely to be in doubt about the immediate human meaning. Our faces carry hearing, smell, sight and taste as well as being associated with intakes of food, drink, and breath. Besides all these obvious sensory needs we 'read' faces as the most necessary part of our natural social skills developed from childhood. The word 'face' has a curious etymology that takes its potential meaning much deeper. The *Oxford English Dictionary* cites *fax, facis*, the Latin for 'torch', as a possible derivation. This implies 'light-emitting' and connects with such phrases as 'her face shone with happiness' and 'his face radiated confidence'. A shining face usually denotes an inner harmony in some degree or other. This directs the memory to the 'higher' references or exceptional circumstances such as the face of Moses becoming too bright for the people to bear — or Jesus's face, even whole body, becoming luminous on the mount.[37] Here spirituality is recorded as a shining of intensity in the face of an inspired being.

Maybe much less in a flower? But how much less if we measure the power and intensity of a flower in comparison to the rest of the plant of which it is representative? Flower faces 'shine' in more than

All three are natural stages in the life of the Moon Daisy.

The Fuchsia is a celebrated hanging flower with beautiful lateral spread to its petals and a bell-like middle with dropping stamens. The buds of this flower are particularly elegantly shaped, characteristically fourfold.

Here two bees visit a Cosmos flower for pollen, and thereby fertilize the Cosmos population in their migration. Bees and the pollinating insects are far more vital to human life than most people realize.

Butterflies both pollinate and glorify our experience of flowers, reminding us of dimensions of beauty we rarely experience.

a metaphorical sense just as beauty becomes an inner radiance in any living being. The beauty of a flower is its penetrating power of attraction. Fertilization is never less than a miracle, an inseparable part of the miracle of life itself. It is we who both debase and fail to appreciate this miracle.[38]

The beauty of the face of a flower reaches far deeper than its connection to the insect world. It is the effect on the higher animal kingdom, ourselves in particular, which goes far beyond a merely material functionality. Flowers can be considered the bearers of brilliant ideas. (What an 'idea' *is* requires a serious study, particularly its original Platonic meaning.) Just consider how many great poems and paintings have originated from experiencing flowers.

This cannot but raise the question of what is beauty and in what way does it permeate both the noumenal as well as the phenomenal world. A mathematician will talk of the beauty of a significant equation — even as Kepler describes the golden section as 'a jewel'. This only goes to demonstrate that our sense of beauty is inextricably connected to our sense of order, harmony and 'rightness' in general. One of the *Oxford English Dictionary*'s definitions of 'beautiful' is 'impressing with charm the intellectual or moral sense, through inherent fitness or grace'. Plotinus recommends us to 'seek Beauty Itself', not merely to be satisfied when we think we have found some*thing* beautiful.

Surely 'purpose' cannot be divorced from the experience of beauty — and purpose itself has material, social, cultural as well as inspirational meaning.

We can, and in this book do, explore one obvious universal language of beauty — that of geometry or the beautiful order of space. This can at least demonstrate that there is more underlying the form of petals and faces of flowers, in terms of symmetry, proportion and harmony, than meets the casual eye. There is also symmetrical colouration on the petals and this includes the veins which in some flowers greatly enhance the symmetry of the petals. Also looking at any flower 'front-on' offers us the most comprehensive single view, although this in no way diminishes the other views which explain other important aspects — particularly how the flower unfurled into being. In the side views we find the spirals of unfolding clearly enacted.

The pre-flower stage of the bud itself is another profound study — as we have already mentioned. In the 'bud stage' we have the opportunity to look at the early expression of the spiral motion that seems to characterize so many flower unfoldings. The spiral in itself is a universal pattern which occurs as much in the largest as well as the smallest images or events in our universe. Life takes time to possess space in spiral form. These range from galactic spin to the decay spirals of elementary particles. It is in proportion that we can 'detect' a special significance. Certain angles determine these and this is where arithmetic and geometry meet. The soul of the world is a self-moving spherical motion according to Plato. The human study of the circulation of the heavens is beneficial to our souls, he says, as it is to be taken as a guide to regulating a *harmonia* between the human and the universal souls.

This flower demonstrates a beautiful fiveness and great subtlety of colouring, often enhanced by delicate veining in the petals.

The Rose is a never-ending wonder of beauty. The spiral unfolding of its fivefold petals, its colouring and its exquisite perfume have overwhelmed many sensitive hearts over its years of association with the human family. However it has been cultivated to have many more petals than its original five.

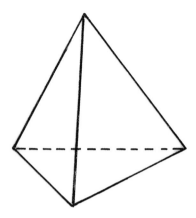

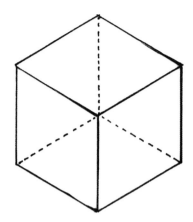

1. TETRAHEDRON

The first solid figure and symbolic of 'fourness' in Plato's *Timaeus* dialogue. It was called the molecule of *fire* and holds layers of meaning not to be reduced to literalness.

4. CUBE

The most familiar of these forms and most logical in its singularity of measure. Symbolic of *earth* for Plato and equally symbolic of the earthly plane for other cultures. Embodies 'order' for most people. Made up of six square faces, eight corners and twelve linear edges.

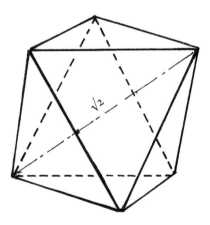

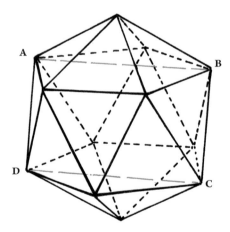

2. OCTAHEDRON

The second solid made from triangles in Plato's *Timaeus* symbolizing *air*. It has eight triangular faces, twelve linear edges and six 'corners' or points.

3. ICOSAHEDRON

The third solid figure in Plato's *Timaeus*, symbolizing *water*. Also made of triangles, this figure 'contains' the golden rectangle between two opposite sides, as indicated by **A,B,C,D**. We find this form in the structure of some viruses.

Archetypal Geometric Form and Spirals

There are three special geometric relations that give rise to spiral expression by increasing in the same ratio at each successive stage. These ratios are inherent in the 'Platonic figures' that Timeaus suggested were the basis of the structure of matter itself: the molecules of earth, water, air and fire. These are now known as the four 'states' of materiality: the solid state, the liquid state, the gaseous state, and the radiant state. We have also affirmed in modern times that the molecular world is 'built' geometrically.

From the primary solid or the tetrahedron symbolizing fire, we find the √3 (square root of three) proportion. This special proportion lies in half of an equilateral triangle. (A 'square root' means the number which when multiplied by itself will give the whole number required. An example is the square root of 9 being 3 as 3x3=9.)

The next primary solid is the octahedron symbolizing air, or eight-faced solid made up of eight equilateral triangles. The diagonal distance *through* the octahedron is the square root of two if the edge of the solid measures one. This is usually rendered as √2.

Finally the last and most powerful 'root' proportion is to be found in the third primary solid, the icosahedron. It is a curious coincidence that life relies upon water for its very existence, and in Plato's cosmogony the 'water' molecule was icosahedron in form. The golden ratio so important to organic life and Plato's atomic molecules meet here. This solid has 20 faces, 12 points and 30 edges. The internal diagonal proportion is called the 'golden ratio' or the square root of five plus one divided by two. This results in the irresolvable golden number which we encountered previously and which can be summarized as 1:1.61803398 . . . etc. (to shorten this we can use 1:1.618 . . .). This special number when multiplied by itself becomes 2.618033989 . . . etc., thus maintaining the identical irresolvable sequence beyond the decimal point. This is a particular mathematical wonder and possibly one of the reasons why Kepler called this proportion a 'jewel'. It is found at all levels and scales in the natural world.

As we have already noted, in the plant world this special ratio is *approached* time and time again in the sequence 1, 1, 2, 3, 5, 8, 13, 21, 34, 55 etc.[39] This internal proportion is a continual wonder and is one of the

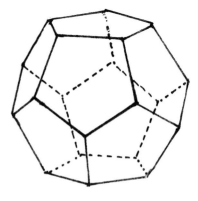

5. DODECAHEDRON

This important culminating member of the set of five totally regular geometrical solid figures was held in special awe by the Pythagoreans, who were reportedly forbidden to divulge its form. Plato also carefully avoided elaborating on it. For our times it was used for the monstrous atom bomb during the Second World War.

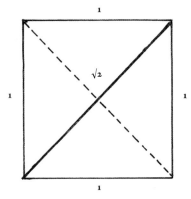

Line-centred view of tetrahedron. An elementary and elemental fact draws attention to the important difference between arithmetic and geometry. The diagonal of a square with sides cannot be expressed in whole numbers but can be drawn with great precision geometrically. The square root of two never reaches a final whole number solution. Plato called it *arretos*, meaning ineffable.

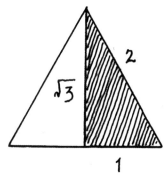

This figure, the equilateral triangle or triangle with equal sides and equal angles (60°), was the very first proposition or theorem of Euclid's Elements, the most influential scientific book for thousands of years.

reasons Plato said that it was 'proportion' that binds the universe. Plato did not openly publish this golden ratio as he was presumably under Pythagorean oath not to reveal it.

We will illustrate how the special molecular proportions of Plato's figures give rise to 'organic' spirals. Naturally there are many others but those inexpressible in whole numbers are the ones which seem to occur with the greatest regularity and were particularly revered in all traditional civilizations in their temple architecture or sanctuaries. This recalls the principle that a temple is a mesocosmos, and the 'gods' are the timeless principles of that cosmos.

This fivefold aspect is especially significant in that the majority of wild flowers are pentagonal or five-petalled by nature.

Studying the orbiting rhythms of the planets has always been a stimulus to human understanding. The luminary Jewish poet of medieval Spain, Solomon Ibn Gabirol, wrote penetratingly on flowers being the reflections of the planets or stars in his poem *Winter with its Ink*.

> Winter with its ink of showers and rain,
> with its pen of lightning and palm of clouds,
> wrote a letter of purple and blue
> over the beds of the garden.
> No artist in his cunning could measure
> his work beside it — and so,
> when earth longed for the sky
> it embroidered the spread of its furrows like stars.[40]

We get a similar inspiration from Rabindranath Tagore, the Indian poet:

> The stars at night are to me the memorials of
> my day's faded flowers.

A true poet or true philosopher (lover of wisdom) will inevitably arrive at the concept of inseparability when meditating on the cosmos. All parts are parts of a single whole eventually. This leads to the depth of meaning of the word *relationship*, which must lie within all things considered by the human mind, as mind itself is one. Whatever else flowers do to humans, they form relationships even if these are unexplainable in words. That there are certain mathematical proportions that appear to inhere within flowers should not be an undue wonder, and their symmetry is open to all.

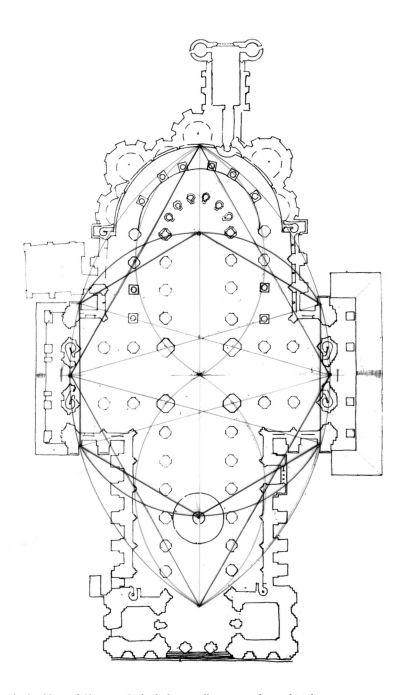

The builders of Chartres Cathedral were all masons educated in the seven ancient
liberal arts and the proportions associated with them. Here we see the influence
of the *vesica piscis* proportioning the whole of this Christian temple. The
Platonic School at Chartres was particularly inspired by Bishop Fulbert with the
encouragement of Gerbert who became Pope Sylvester II.

Plato reminds us in his cosmogony *The Timaeus* how important seeing the night sky was for humanity. These points of light remind us of the laws of number and objective order.

Dandelions and Daisies are usually present in any cultivated lawn whether invited or not. They have been used by many poets, particularly Ibn Gabirol, to represent the stars of the night sky.

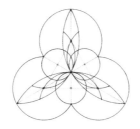

THREE

THE GEOMETRY OF FLOWERS

But the creative principle resides in mathematics.
In a certain sense, therefore, I hold it true that
pure thought can grasp reality as the ancients dreamed.

Albert Einstein

Here Kuan Yin, the Goddess of Mercy for Buddhists, displays
her conventionally-shaped light body which is also the confirmation of her true being.

The Mystery of Flowers,
Their Chosen Symmetry and
Their Intrinsic Geometry

The centre is still and silent in the heart of an eternal dance of circles.
                                                            Rabindranath Tagore

Near your breast bone there is an open flower,
Drink the honey that is all around that flower.

                                                            Kabir

We must find out for ourselves what is the relationship between nature
and each one of us.

                                                            J. Krishnamurti

The single-petalled Arum flower (*Calla palustris* or *Zantedeschia*) has a classical dignity and a curious similarity to images of the Lord Buddha and his light-body reality. Another version of this flower is called 'Lords and Ladies'.

We cannot be certain what gives us the appetite for order, but we can be certain that it was there from the beginning — whatever cosmogony we ascribe to. We all appreciate an ordered life with not too many surprises. Life is not possible without order. Life has to be ordered so as to be spontaneous.

Symmetry must rank highly as one of the chief mysteries in this impulse for order. The symmetrical arrangement of elements within the same system tends towards balance, which *is* symmetry. The human body has a high degree of natural symmetry which is expressed largely in a twofold way — in common with most of the animal kingdom. From this we get our right-handedness and left-handedness which maintains our natural bodily balance in movement. We know our internal organs differ yet they are distributed to maintain balance and an interior harmony. Flowers express a plethora of beautiful symmetries ranging from the twofold to the manifold. The most predominant symmetry, particularly in wild flowers, is fivefold. It is intrinsic to this pentagonal symmetry that the 'golden' proportion of 1:1.618 etc. is present as we have consistently noted. Fruit blossoms are most often fivefold, a symmetry that is powerfully visible when we cut an apple across its meridian — passing through the five enclosures for the 'pips' or seeds.

There is no better demonstration of sameness and otherness than in a great bed of tulips (this is in Hyde Park, London). They are all the same variety of white Tulip and in a way identical, yet each is clearly slightly different from all its neighbours. The whole is 'a bed of white Tulips'.

Symmetry is described in the *Shorter Oxford English Dictionary* as: '1. Beauty resulting from right proportion between the parts of a body or any whole, balance, congruity, harmony. 2. Such structure as allows of an object being divided by a point or line or plane or radiating lines or planes into two or more parts exactly similar in size and shape and in position relative to the dividing plane, etc. The repetition of exactly similar parts'. It is an interesting fact that we all find beauty in symmetry — yet this is not easily explained. Maybe common sense says a balanced arrangement is least threatening to our visual understanding? Whatever else, flowers most evidently 'speak' to us and affect us in their radial symmetries.

Firstly we have to establish that 'shape' inevitably means *some* sort of geometry (or stereometry) classified so adequately by Euclid within the Platonic tradition. Geometry in this heritage was seen as the bridge between our world and the higher world of causation. Curved geometry was considered the 'higher', or of the heavenly order, whereas straight-line geometry was considered representative of this world. The curve represented the transcendent, the straight the rational, the calculable.

All petals when 'complete' go through a straight-line direction as well as an overall 'curved' unfolding. Both are embraced by symmetry. Thus the universal order is re-enacted by every flower.

There is more than a single way to measure the geometry of the petals and flowers. Not only does each petal have its own characteristic profile and curvature but the ensemble of all the petals is what we call the flower. This collective geometry includes the total symmetry.

Whatever else we might conclude about symmetry it is valuable to acknowledge Plato's first principle of the triad of Sameness, Otherness and Being. In terms of flowers this means that no two petals, let alone leaves, are ever *exactly* the same in a single bloom yet they are usually clearly identifiable as an Oak, Ash or Willow. In short all scrutiny of the natural world will entail Sameness and Otherness. These two qualities will always seem antagonistic or contrary and therefore will never be resolved until our acknowledgement of Being is recognized. Being reminds us of the extraordinary balance of nature and balance of the universe itself. Through cognizing Being we learn to see Sameness and Otherness *simultaneously*. In short we can experience a host of Tulips though common experience reveals that each will be slightly different: each is unique yet unquestionably a *Tulip*. Paradox alone can achieve unity.

Here we are taking a further Timaean ethos of working within the

TWONESS
Two flowers and two leaf-arrangements demonstrate their adoption of the use of twoness.

THREENESS

These six examples of flowers adopting the strategy of threeness for their petals
demonstrate difference in sameness. Three of these are Irises.

boundaries of likelihood or seeking the most likely underlying order. As we live in the world of necessary change, we cannot expect to have available to our unreliable senses 'absolutes' or 'perfect' examples in nature to study. The best we can hope for is to find the 'most likely' indications of the originating perfections, as Timaeus insists at the beginning of his Cosmogony myth. These we consider valid to call evidence of 'archetypes' or the determining principles. The existent flowers echo or reflect the archetype or originating paradigms. These are enshrined physically in what we can call their genetic memory bank and their reactions to their immediate environment — sunlight, for example.

When seeking a solution to the curves of any particular flower petal we can only choose the most likely or closest curve. This requires a huge effort to exclude any pre-formed theories, shapes, characteristics and so on. So we rely, as objectively as we can, on the 'nearest solution' to each petal curve. This is the 'gentle empiricism' of Goethe determined by an internal integrity. We must not 'fool' anyone least of all ourselves in seeking the underlying principles of geometries. To make the utmost effort to find the nearest to the truth was Timaeus's advice. Solutions can only be 'offerings' not proofs. Proofs are after all only based on a faith in an accepted 'rule' or conventional principle.

To evoke the 'gentle empiricism' of Goethe, we might first observe the value of Goethe's theory that petals are transformations of leaves.

We have the one-petalled Aspidistra flower with its central stalk. From the imaginative view point this can be seen as a remarkable likeness to the lovingly-represented light body that surrounds ancient Chinese and Japanese representations of the Lord Buddha. From this singular white petal and other coloured petals, the next occurrence is of dual-petalled flowers. These are either with only two petals or pairs of two petals making up the flower. Two-petalled flowers are rare but can be found.

A triple seedpod. What profound economy and expressiveness of form.

From here we move to the much more frequent three-petalled flowers such as the Snowdrop, the Tulip, the Iris and the Lily. So many apparently sixfold flowers are really a double three in their petal structure under closer scrutiny.

The threeness of a leaf or flower immediately indicates a sixness in the handedness of the leaf or the arrays of the petals. However, the distribution of threeness in the Lily and Iris is beautifully evident.

Next there are some very beautiful fourfold flowers or flowers with balanced spreads of four petals, not least the Clematis, the Balloon Flower and the beautifully fragrant Wallflower. The Clematis is a good example, but this flower also unfolds beautifully in a double four (or eight-petalled) form.

Next we come to the most frequently occurring symmetry in natural wild flowers: the fivefold or pentagonal symmetry. The list is impressive and includes the original Dog Rose (that Rose form which most other Roses seem to have been bred from), the Buttercup, the Herb Robert, the Periwinkle, Borage and so on. These all display fivefold symmetry so beautifully yet so individually and at quite different scales.

What becomes apparent once we get to the fivefold symmetry is that the 'Golden Progression' is underlying the sequence. This organic progression is known as the Fibonacci sequence — after the Italian mathematician who it is believed was tutored in his early days by Muslims in North Africa where his father was an Italian provincial governor. As we have previously noted, this sequence may be called 'golden' as it progressively comes closer to the 'golden ratio' of 1:1.6180339 . . . This ratio is arithmetically *in*expressible in whole numbers, like *pi* (π), yet can be accurately expressed *geometrically* — particularly in the pentagon with its diagonals making up a pentacle. This remains a wonderful mystery to the human mind.

Sixness is found in the Daffodil whose petals fuse into its hexagonal shaft.

Next we come to the most rare of flower symmetries, the 'virgin' seven. This does occur in such as the Primula flower yet is never completely balanced in its symmetry. It also occurs occasionally in the Tulip — once again revealing an imbalance, yet exquisitely beautiful.

Next we have eightfold symmetry which is most beautifully represented by the Cosmos flower in a range and variety of colours. The sacred Lotus is in this symmetry and the Clematis is another fine example.

Ninefold symmetry does occur although rarely: we find it in certain Cactus plants and in others that are exceptions rather than the rule. Again, nine is a notoriously mysterious symmetry. From the geometer's viewpoint this is because it cannot be constructed *exactly* with a compass and straight edge. Good approximations can however be achieved.

Tenfold symmetry is also found occasionally, most particularly in the Passion Flower. Yet even here it seems to be composed of two fives. This same flower will also occasionally exhibit an eightfold symmetry. Ten, we recall, has the golden ratio inherent in its geometric form.

Elevenfold symmetry is *very* rare but does occasionally occur in flowers of the Daisy family (*Compositae / Asteraceae*).

Twelvefold symmetry also occurs in the Daisy family of flowers but is not common.

Thirteenfold symmetry is more common in the multi-petalled flowers like the Daisy family. But as this is one of the numbers of the *golden sequence* it is not at all surprising.

Another example of four-petalled flowers — the beautifully perfumed Wallflower.

FOURNESS

Here again we show a set of six flower types all utilizing fourness as their petal strategy.
What might 'fourness' be saying to the insect world?

FIVENESS
This symmetry is both the most 'friendly' and most common among British wild flowers.
What immense variety in sameness of symmetry!

Twenty-one, the next in the *golden sequence* is found again in the Daisy family and confirms the underlying golden order: 5, 8, 13, 21.

However there are many exceptions in flowers with multiple petals as there are a bewildering number of symmetries in the heads or seedpods of poppy flowers — even though the petals are usually fourfold and on occasions sixfold.

One of the questions that arise when looking at petal symmetries is: how much have humans 'engineered' the numbers of petals? This raises the further issue of how compliant the 'natural' or 'original' flower is, or was, to being engineered into a more complex form —let alone into quite different colours. In the author's view, Tulips are the most painful examples of human interference, with the obviously unnatural feathering of the petal fringes. However there is also a staggeringly beautiful range of possibilities that have come out of human breeding, known as cultivars.

One of the reasons why the author enjoys very small flowers such as the Herb Robert is that the visual effect is so powerful while the size is not much bigger than a human thumbnail! Fortunately the smaller more modest wild flowers have not attracted so much attention of flower 'breeders' or 'engineers'. However, there is currently a growing revival of interest in the value of wild flowers in their natural setting. Durlston Head in Dorset is one such important site receiving a growing number of wild flower enthusiasts from southern England.

SIXNESS
A display of nine sixfold flowers, some of which we have already considered.
The range of choice of expression within sixness is sublime.

SEVENNESS
A particular variety in petal arrangements.
The bottom-right photograph actually shows seven flowers arriving simultaneously in a cactus.

EIGHTNESS
Four good examples of eight-petalled flowers.
We might have added the Lotus but this will come later.

NINE, TEN, ELEVEN, TWENTY-ONE
Counting petals is a beginning to understanding the arithmetical nature of flower petal choice.
Names are important but here we wish to focus on other aspects of the flowers.

Here we see a Camellia with multiple petals progressively
approaching the sphere. The petal-sizes reveal the
mathematical proportional system.

Petal Geometries and the Principle of Radiance

In this book we have deliberately chosen those flower petal arrangements that radiate fully and thereby respond to the most universal geometry of complete rotational symmetry. This means that each flower we have focused on and analysed responds to the regular subdivisions of an overall encircling form. This is not to ignore the many other types of flowers that are not radially symmetrical. It was the philosopher Proclus, the last great Platonic academician, who reminded us that the circle, or circular, is the language of the *intelligible* or 'heavenly' domain while the orthogonal or straight is that of the *sensible* domain. The circle indicates the archetypal source — significantly in the form of the sun itself.

A circle naturally symbolizes a totality of space. It can also be seen as the circular flow of time, each point of symmetry marking an interval of time, like the hourly numbers and second marks on a clock face. (When the ancients measured time and angles in minutes and seconds they were acknowledging the complementarity of time and space.) The circle paradoxically is the natural expression of wholeness and oneness as well as the representative of the 'zero'. The most inexplicable conclusions derived from our analyses were the positioning of the points, usually on the embracing circle, from which the curvature of the petals could be struck as radii. A fine example of this migration of the centres from what would be considered natural comes from the careful analysis of the curvature of the six-pointed or six-petalled flower. The most familiar and naturally occurring pattern struck by a pair of compasses when tracing the curves, without changing the radius between the point and the scribing edge, is the six-petalled geometrical 'flower'. It is a fine example of profundity in *simplicity*. Sixness is the intrinsic nature of the movement (life) of a pair of compasses. The sixness of 'creation' is adequately symbolized in this most simple of geometrical exercises.

1. First we take a comfortable, clean sheet of white paper supported evenly by a smooth surface, and sufficiently fixed to its supporting surface so that it does not slip. Then we take our compasses and estimate the mid point of this piece of white paper. This is done by first holding the top joint of the compass with the natural writing hand. The other hand has the equally important job of steadying the static point and directing it into the required position for this drawing: the centre of the page on which we intend to draw the pattern. Finally we estimate a convenient radius to give our first circle plenty of room about it. Once the first circle has been drawn, this circle acts as the standard guide for all the further arcs to be drawn. It is so easy to draw with compasses that its profundity is too easily overlooked.

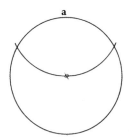

2. Next we move our compass point and guide it into the topmost point of our first circle (at **a**) and draw a second arc of the same radius as the first. This second arc gives a clue as to how we find the centres of two more arcs.

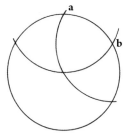

3. The third arc is centred on the crossing-over point of the first and second circle (at **b**)— but here on the right side. There is another crossing over point on the left side, which we will return to.

This process, of using the intersecting points of the similar circles as they cross one another, is the way in which we proceed to produce the flower form within the circle.

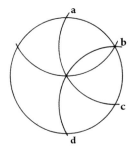

4. By the time we get to the third similar arc (centre point at **c**), we will have created the first of the 'petals' within the first central guiding circle.

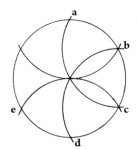

5. Next we place our compass point on the lowest intersection, which we have marked **d**, and draw our fifth arc. This will create our second 'petal' below the first inside our guiding circle.

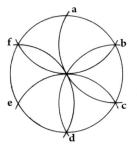

6. The next circle will be centred on point **e** which is the intersection made by the circle centred on point **d** and the central circle. Now we will have created three 'petals' of the eventual 'flower' within our original guiding circle.

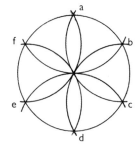

7. Finally we find the centre point of our last similar arc, located on two existing intersection points, **e** and an earlier point from our first circle. Now by scribing our seventh similar arc we will have created a six-petal flower within our original guiding circle.

The Primary Geometrical Flower

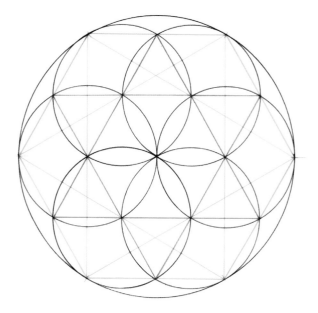

The beautiful simplicity and profundity of the sixness of a circle divided by its own radius. This exactness of six equal divisions of the containing circle makes a profound comparison with the numerical value of pi (π) which cannot be resolved to a whole number — whereas the geometrical solution is precise and beautiful.

Six is traditionally a perfect number as $1+2+3 = 6$ and $1 \times 2 \times 3 = 6$. Also six are the days of creation in the three monotheistic faiths and six is the natural division of a circle by its own radius. Above we see the flowering sphere of the onion plant.

Our illustration shows a completed geometric archetype, profound in its simplicity, which we will relate to the occurrence in an actual flower. This pattern makes a close representation of the flower form in the illustration *yet it does not do so exactly*. In fact there is a minor shift of the central points of the defining area of the petals which, when studied, are situated on lines connecting different points on the peripheral guiding circle. This outer circle is the one we referred to earlier and is the 'controlling' periphery within which the petals can be constructed. The points quite clearly are not defined by *material* or actual positions of the resultant petals — rather they are within the 'field' *around* the petal forms.

It is just this invisibility and intangibility of the static controlling points of the petal-defining arcs that reminds us of the concept of morphogenetic fields. However, certain contemporary botanists are mostly concerned about the petal being determined by the genes of the flower. Our studies show that there is also an unexplained outer constructional geometry, which indicates an 'outer' form-controlling aspect. This could be called the geometric 'field'. The exact nature of this 'field' is yet to be explained.[41]

The Tulip, an age-old favourite of the human family. Face-on views demonstrate
the power of its symmetries (six, seven and eightfold above). Although thoroughly 'used' by
cultivators this flower still holds its beauty significantly. Its history is a dramatic tale.
a. This Tulip unfolds in a remarkable triangular fashion. **b.** Sevenfold.
c. The much less usual eightfold Tulip flower.

The Search for the Controlling Archetype in Flower Geometry

You are filled with ever fresh wonder
When flowers unfold on their stalks
Above the leaves' multiform framework.
In their glory however, they herald a whole new creation.
Yes, these pure-coloured petals feel the touch of God's hand.

Goethe

The difference between the geometry of flowers and the flowers of geometry is the same as the difference between the absolute or archetypal and the actual or apparent. In the Greek tradition these two views were characterized by the *Intelligible* and the *Sensible* as we have already explained.[42]

These two 'worlds' have also been called reflections of each other. In fact the relationship between principles and material actualities has engaged philosophical minds throughout the lifetime of recorded human thought. It is directly parallel with the equally universal acknowledgment of spirit and matter — the intangible source and the material expression. (Today the conventional scientific community chooses to call such principles the 'laws' of nature.)

Plato is conventionally associated with the Archetypes and his student Aristotle with a focus on the categorization of Actualities. This is dangerously divisive if taken too simply, yet the emphasis of each of these vastly influential thinkers does tend towards these extremes. But another intellect, Master Chu (Chu Hsi) of China who devoted much thought to the relationship between *li* (principle) and matter (energy force), came to the conclusion that they are virtually interdependent. 'But', he added 'I would have to make principle slightly prior'. Plato puts soul *definitely* prior to the bodily manifestation of the Universe in the *Timaeus* Cosmogony.[43]

Traditional wisdom, it seems, has always placed both within the same unity: matter cannot be appreciated without principle nor principle without matter.

When we take a single image of a flower we know it will exhibit clear specific characteristics of its family and species yet will also inevitably carry

One of China's most influential philosophers of the Middle Period. Master Chu cultivated common sense and spoke on worldly matters as much as philosophical abstraction. His works (wonderfully translated by Wing Tsit Chan) are highly recommended.

the imperfections or eccentricities of its individuality.

The search for principles underlies most if not all authentic 'scientific' experiments. What conclusions we come to will be a matter of each person's 'sufficient' reasoning or 'faith' in the conclusions. This means the explanation that is 'most like' the intrinsic or inherent truth underlying our experience or calculations will be the most acceptable. This, in the opinion of the author, is one of Plato's major contributions to the Perennial Philosophy — that of a unified definition uniting the actual or sensible with the theoretical or ideal through 'Being'. ('Being' inevitably having to embrace paradox.)

In our presentation we will first look at how the geometrical and symmetrical forms appear to be expressed in the natural world as good honest likelihoods. This will be compared to the contrasting 'absolute-reflecting' pure geometry produced by compass and square or straight edge. The subtle differences are as instructive as the philosophical issue itself.

The inner governing archetype of a flower is the mystery of its species while the flowers of geometry are the 'timeless' forms of pure geometry. These can be provisionally expressed as indicating absolutes in geometrical drawings. The only tools required are a flat surface (a piece of flat supported paper), a pair of compasses with a scribing (or marking) point and a straight edge or ruler. (Both instruments can be replaced by rope, cord or knotted thread, but this takes far more skill and practice and is usually used at a larger scale.) The language is in points, lines and surfaces or shapes. They are remarkably self-making or self-proving when drawn carefully, particularly when drawn by the human hand in an act of reverence and respect. The ancients described such an activity as 'participation'. The higher the goal the deeper the understanding.

At this point it is important that we acknowledge the difference between the mechanically-drawn geometric diagrams and hand-drawn versions. The difference is not merely a matter of human debate but rather a matter of attitude and sense of what the act and results ultimately *mean*. Since the industrial revolution our mental habits have been conditioned into only thinking of a mechanical solution (an attitude assuming that 'science' only leads to 'technology'). This attitude has engaged, even overwhelmed, the scientific community as much as the general public. Geometry no less than other arts can be experienced on all four levels that Socrates outlines in the 'Divided Line' — practically, symbolically and socially, culturally and inspirationally. Geometry is not a mechanical craft. It is a tool according to Socrates by which to experience the 'ever true' and to lift one's consciousness into the intelligible realm of pure light.[44] The clearest indication of the inappropriateness of computer graphics or geometry drawn by a computer is the singular loss of the centre-point of a circle. This essential guiding centre-point, which is an intrinsic necessity when using hand-compasses, represents the heart of the circle. It represents the symbolic mystery of the 'hidden' creator. Even the godless theorizing which offers the Big Bang as the creation principle is said to begin at a 'point singularity' with *nothing* prior to it. So the point has become one of the most powerful of mysteries for all. What do we mean by 'getting to the point'? If there is no point what else do we have but nihilism or meaninglessness — the most psychologically destructive of all human concepts? Wonderment and mystery, however, inspire and stimulate — for the better.

'He who sees the Infinite in all things sees God. He who sees the Ratio only sees himself only' — William Blake.

An extraordinarily inspired insight into the true nature of geometry by the British visionary. (See Kathleen Raine's *Blake and Tradition*, Routledge 2002.)

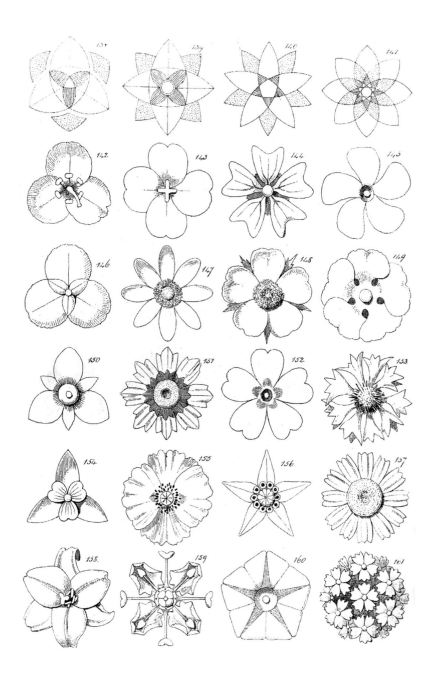

These flower drawings, no doubt based on good observation, were probably
directed toward educating British designers at the turn of the last century.
Drawn anonymously, they have much to offer.

An Introduction to the Geometry of Flowers: The Primary Symmetries

As Plato made clear in his cosmogony, *The Timaeus*, there are three Primary Symmetries that control distribution in space. As flowers need to take possession of space it is inevitable that they will express these fundamental patterns. Contemporary science does not disagree with the principle of essential inevitable symmetries.

These three regular patterns of distribution are from a regular three-ness, a regular fourness or a regular fiveness. Which we can call the triangular, square and pentagonal.

Pythagoras was immortalized for my generation through leaving his name on the proposition that the square on the hypotenuse, the longer side of a right triangle, is equal to the squares on the other two sides. This truth is also immortalized by the right-angled triangle that has sides of three units, four units and five units. An immensely practical fact that is as much in use today as it has ever been; used by craftsmen or builders who need to establish a right angle in what they are building or laying out.

These three numbers take on more than a practical aspect as they are clad in deeper symbolism — especially by the Pythagoreans and by Plato. Number is, after all, the primary abstract science in which we can all engage, safely and in the knowledge that everyone consents to its nature and behaviour. However the symbolic level will differ according to the culture civilization or inspiration.

The following analyses of the plant world arise from taking the most simple shapes — which are equally archetypes — and looking into the logic (or transcendence) of how the plant world has adopted them.

It was no doubt the wisdom tradition brought by Pythagoras from the ancient Egyptians and confirmed by Plato during his sojourn in Egypt. This is beautifully embodied in *The Timaeus* dialogue of Plato. It is also the springboard for the Aristotelian corpus — even if he placed a greater emphasis upon the impressions we receive from the manifest world. It is good to remember that Aristotle was the student of Plato for at least 18 years, even if he didn't fully accept the doctrine of 'ideas'.

The fundamental molecules of materiality from which all things are made was offered by Timaeus as having been based on the most elementary and elemental geometry.

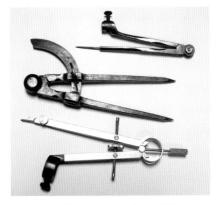

The tools of the geometer from different periods. A schoolchild's serviceable instrument, a mason's tool and a modern draughtsman's tool — all simple compasses.

The half-equilateral triangle, Plato's 'most beautiful' triangle. A modern 'plastic' version.

Chlorophyll, so essential in the conversion of light into nourishment in a plant, has two beautifully symmetrical molecules.

These three shapes or symmetries are the threefold or triangle, the fourfold or square, and the fivefold or pentagon.

Modern 'material' science has demonstrated that the 'atomic' theories of the Greeks, especially Timaeus, have been confirmed. Also that the molecules that are utilized by life have been dramatically uncovered – not least the DNA molecule. Our analysis shows it too, as Plato predicts in *The Timaeus*, is founded on proportional principles. Proportions that reflect through all levels of our universe. Proportion in itself, like symmetry, is intangible. As Timaeus suggests — the invisible soul is fundamentally proportion and thereby itself intangible. The basis or highest universal law we can all engage in is number, no science is verifiable without number, yet *qualitative* as well as quantitative number.

Three is the first number that can make a shape, e.g. three points and three lines enclose the first shape — the triangle.

Four is the number of worldly order in the spatial directions, the states of matter but, more importantly, for the Pythagorean philosophy, four-ness is the Tetraktys. This means, one plus two plus three plus four equals ten, a symbol of considerate depth — venerated by the Pythagoreans.

Five is often called the number of life as it so beautifully embodies the golden proportion. As the centre of fourness geometrically it takes up the controlling position being equally related to all four corners of the square. It represented 'spirit' or the animating principle at the heart of matter. Thus it was called the Quintessence.

Flower analysis can be conducted in many ways. We have chosen to take the Platonic or traditional way. This means using the most simple and cardinal numbers, figures and symmetries. Keeping in mind at all times the profound wholeness and unity of our Universe — as Plato advises us to do.

This criteria will naturally point us in the direction of starting at 'one'. This can only be responsibly represented geometrically as a circle, yet this does not eliminate such as a simple straight line, a simple point, or a single plane. Fortunately the circle 'contains' a single point enclosed by a single perimeter.

From the single circle we can follow the procedure that we have before and explore what values or proportions arise from two similar circles, three similar circles four similar circles and so on. The number of petals of the flowers we analyse reflect this principle. That is a wholeness containing a balanced series of sequential numbers and their interrelated forms.

We look at the rare flowers with a single petal. The Arum is such, as is the similar flower that my local flower seller calls the 'Peace Lily'. We follow this by examples of flowers with only two petals — yet in pairs. This then brings us to three petals and we enter the beautiful world of Irises and certain water-loving flowers, the Plantain for instance. Also spring's earliest, the Snowdrop, and so on.

From three petals and the triangle we move to the square of four petalled flowers. The Wallflower being an excellent example. There are many in the tiny Wallflower category from the abundant 'wild' flower source.

From four we move to five-petalled flowers. Of these there are a multitude. Five seems to be one of nature's visual strategies. Numerically this aligns to the golden sequence of numbers — which is so prolific in the natural world.

From five we move to the sixness of petals of the Daffodil, the Crocus, the Lily and so on. We eventually proceed to the sacred Lotus which is eightfold — and linked to this we show how the Poppy seedpod has even more elaborate symmetries in its radii and diameters.

Naturally certain key geometrical proportional diagrams recur amongst the symmetries.

What is so wonderful is how these archetypes are used by the floral world to achieve such immense variety.

We have had to change the scale or size of many flowers so as to be able to clearly analyse them. So remember some of these, particularly the wild ones are tiny — yet powerfully beautiful.

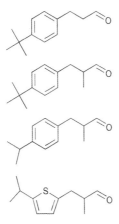

The Lily of the Valley (*Convallaria majalis*), which figures in the portraits of Erasmus the philosopher and Copernicus the astronomer, has had its inspiring perfume redesigned into a number of similar smelling molecules by the 'perfume scientists'. Here they are. What we do to nature for financial profit!

An analysis of the section through the DNA helix.

An analysis of the inner portion of the section through the DNA helix.

The DNA molecule analysed through its cross-section. It reveals golden
proportions within golden proportions in its tenfold symmetry. This pattern
reverberates throughout both macrocosms and microcosms.
See Scott Olsen's book *The Golden Section*, Wooden Books.

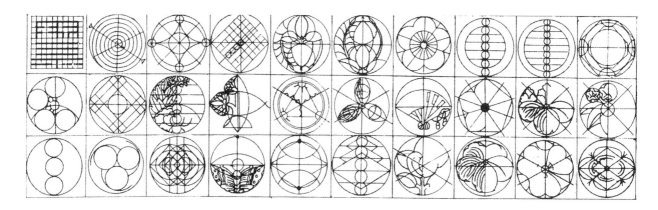

Above we see examples of traditional Japanese crests and sword-handle designs
demonstrating the clear understanding of the geometry of natural form.

Images from Villard de Honnecourt's notebook. An insight into the 'inner
geometry' of the art of Christendom, no doubt influenced by the Platonic
insistence on geometry.

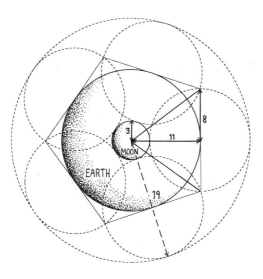

This is the work of contemporary astrophilosopher Robin Heath. The profound simplicity of this discovery is totally disarming. Also the 'flower' factor in the geometry cannot be missed.

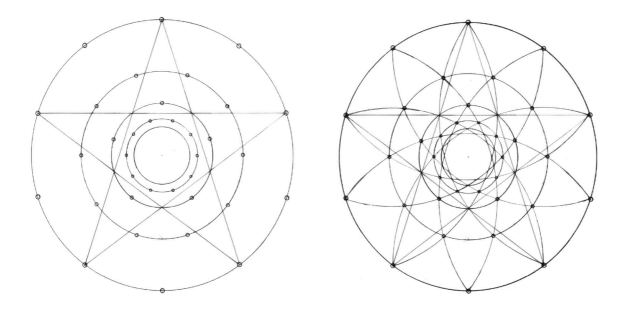

Here we present the significant archetype of five, or the pentagonal symmetry, in regular form. This lies within so many British wild flowers.

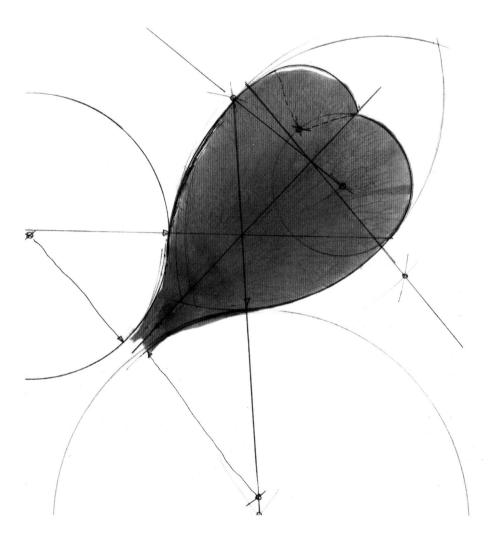

This preliminary analysis of a petal shows how it relates to a basic *vesica* form,
but we have also sought out the radii of the two lobes of the petal. These are
never an exact reflection of each other.

This collection of petals from a Camellia flower demonstrates very well the principle of sameness and otherness. All petals are from the same plant yet it is clear that their separate geometries will be subtly different.

A. PETAL   B. LEAF

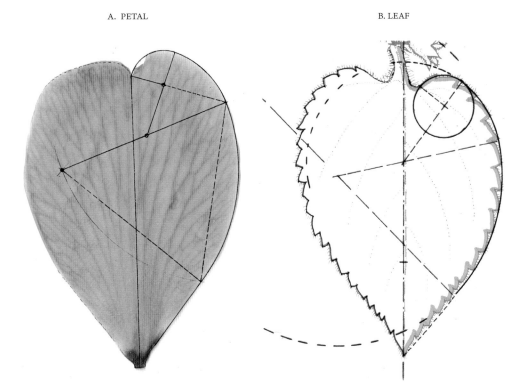

Here we have taken a simple petal to find out the nearest arcs that describe its profile. Remarkably, a leaf has a very similar, but reversed, profile of geometry: the stalk is at the opposite end. This is not easy to explain but is geometrically precise in both cases. Goethe's theory of leaf transformation comes to mind.

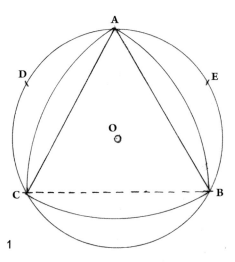

1

In drawing **1**, we have first drawn a circle radius **OA** or **OB** or **OC**. The dashed line **BC** signifies the width of the petal at its maximum.

With radius **OA** we now mark off **AE**, **AD** at this radius.

Next we mark off **DC** from **D** and **EB** from **E** by the same radius.

Now we can draw an accurate equilateral triangle with radius **AC** and **AB** from **A**.

The dashed line completes the triangle **ABC.**

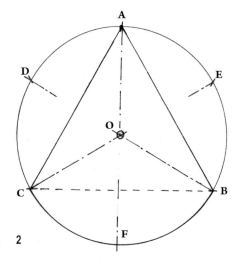

2

To confirm the exact position of centre point **O** we connect **BD** and **CE** with central lines.

Now we can see how the petal geometry responds to the shape based on the equilateral triangle **ABC**. But returning back to the centre point **O** we can now include the 1/3 arc from **B** to **C** of the original outlining circle.

So we have **ABFCA** as the petal shape.

Here we take another group of petals as they have fallen from the flower, and show a very simple archetypal geometry to them. Euclid's first proposition is more than a mere geometric exercise. It is Trinitarian in the fullest symbolic sense. We are reminded of Sameness, Otherness and Being.

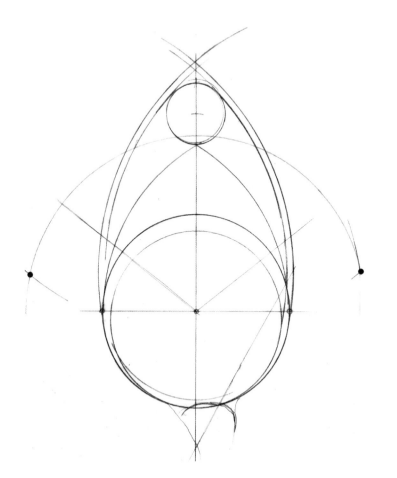

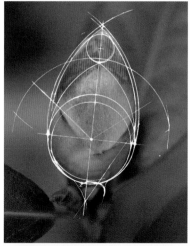

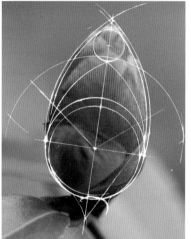

The bud of any plant's leaf or flower will
always signify the new form breaking out
and upward to the light.
The base of any bud tends to be spherical.
It opens like a Gothic arch.

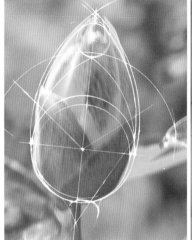

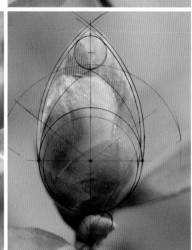

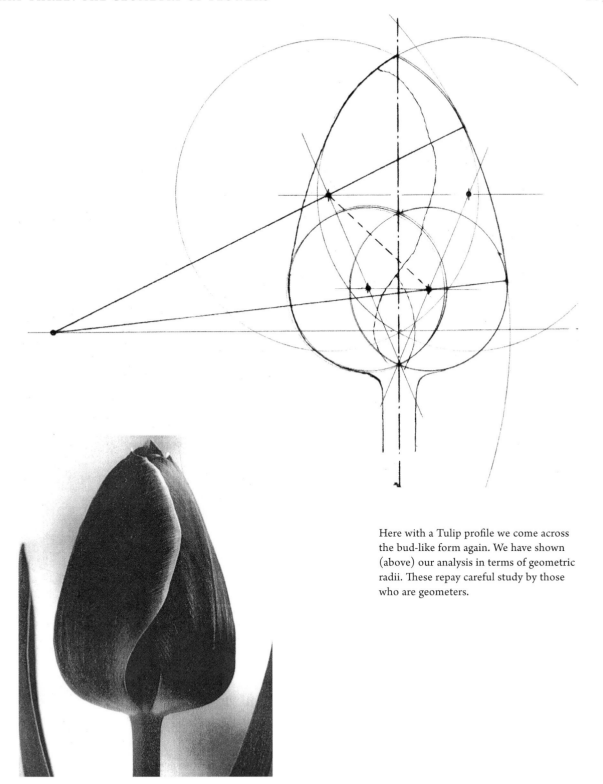

Here with a Tulip profile we come across the bud-like form again. We have shown (above) our analysis in terms of geometric radii. These repay careful study by those who are geometers.

The Arum flower is called a 'Peace Lily' by my local flower-seller. In the top image we see the obvious likeness to the 'light body' of the Buddha (the seedpod being symbolic of his human body). The 'light body' is a *vesica* in essence, exactly like that of Christ on the west portal of Chartres Cathedral. In the bottom image the same flower takes up a geometry very similar to that of a bud — upside down in this instance. Its spiral unfolding is particularly marked.

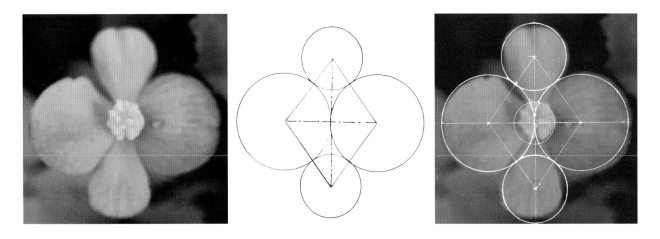

Left. The original image of the flower.   Right. The essence of the proportions between the petal centres as they are determined by the geometry (centre).

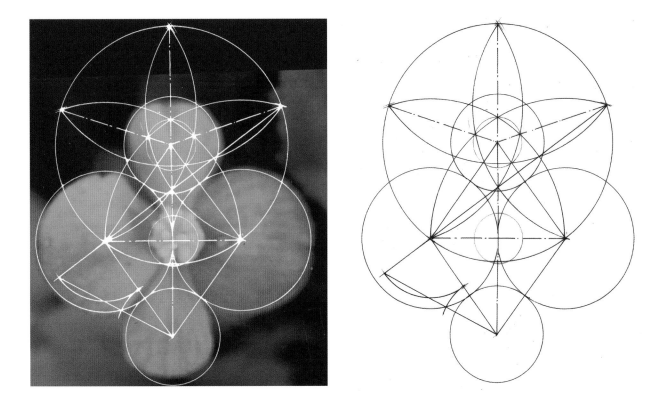

Left. The proportions between the pairs of dual petals in this flower appear convincingly to be founded in the five-petalled geometry.   Right. The golden proportions are evident here, as a careful study of this analysis will reveal.

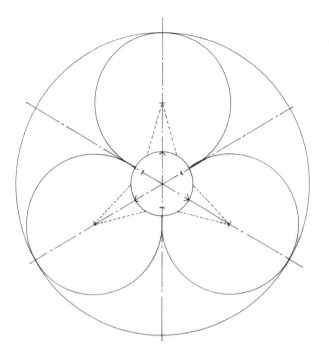

The Iris is most associated with threefold and sixfold symmetry in its
petals and anthers. Here we show a particularly triangular unfolding
and the associated geometry.

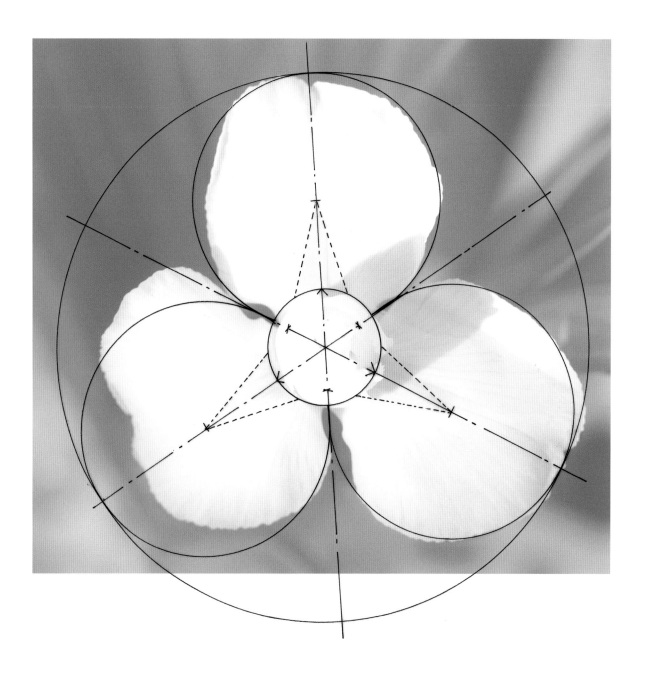

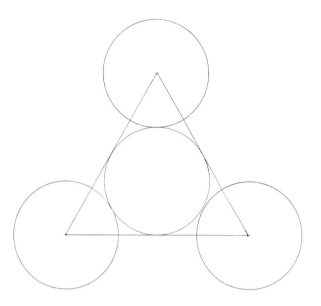

As noted, the Iris is dedicated in the European painting tradition to the Virgin Mary due, it is said, to its Trinitarian nature. It is a remarkable pattern of great simplicity in its proportions. It also relates to the mystery of $\sqrt{3}$.

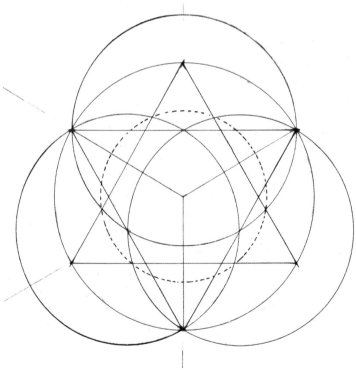

The analysis of this *Tradescantia* has a remarkable similarity to the
structural way three soap bubbles relate on a flat surface. The threefold
symmetry is beautifully precise.

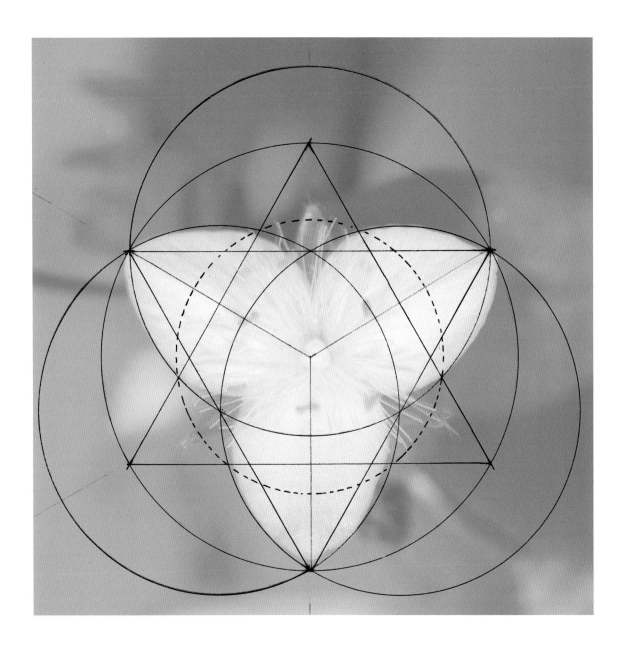

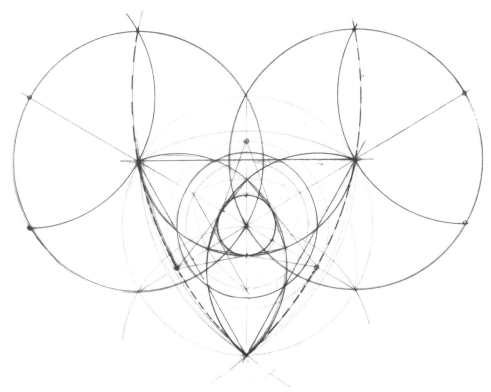

This Tulip opens in a way that demonstrates its threefold and sixfold nature.
The central hexagon is quite symmetrical. The outer threeness of the petals
responds well to analysis as shown above.

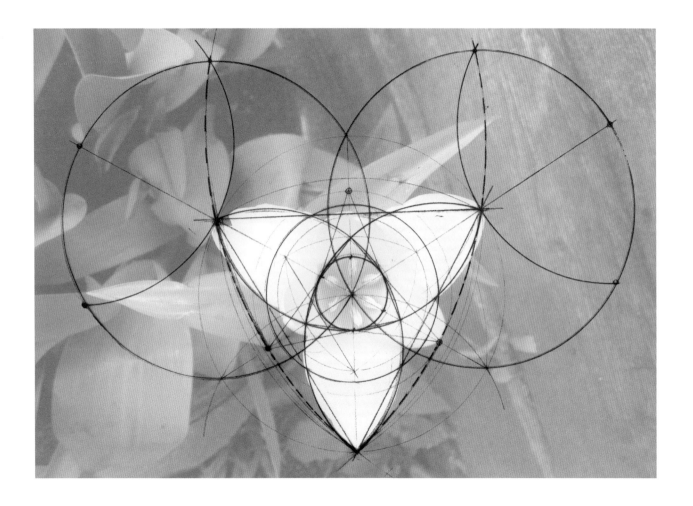

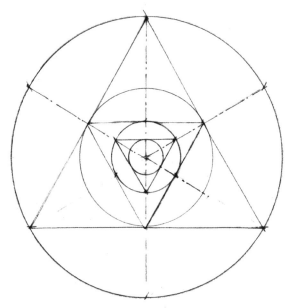

This *Osteospermum* 'Whirligig' flower in the Bellis family has crinkled petals — and they curl on exactly a √3 proportion to one another. This can hardly be treated as accidental. The result is that we have a set of √3 proportions represented as circles within circles. This could be called a 'classical' √3 proportioning series. The coincidences with the parent flower are far too significant to be ignored. The petals are obviously beautiful and their proportions work in a subtle way.

a, b, c, d. The four Begonias above are good examples of twonesses in pairs.
e,f. With its powerfully nostalgic perfume, the Wallflower was a vivid companion
during my childhood love of flowers.
Here we see its relation to four overlapping circles quite clearly.

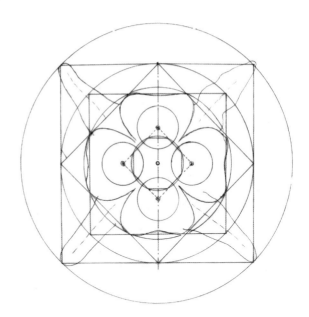

This *Aquilegia* flower has, rather
unusually, a fourfold symmetry which
we have analysed geometrically.

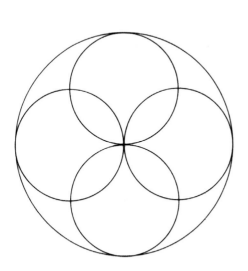

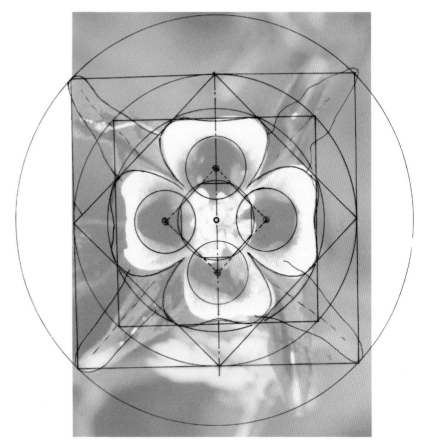

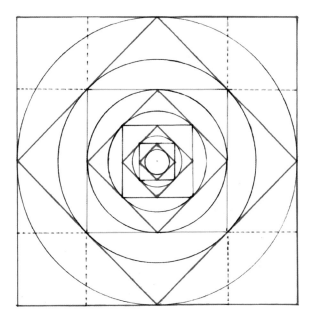

A rare version of this Cistus flower in fourfold symmetry. It is most usually in five-petalled form. In this other chosen symmetry, however, it responds remarkably well to a diminishing or expanding √2 series of proportioned squares within squares.

This proportioning system of squares within squares was
diligently employed by the gothic master masons.

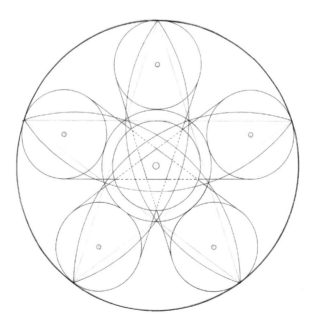

The geometry of the Herb Robert flower has a 'classical' quality reflecting the
same proportions that we have found in so many important places.
Careful scrutiny will be of value for the geometrician.

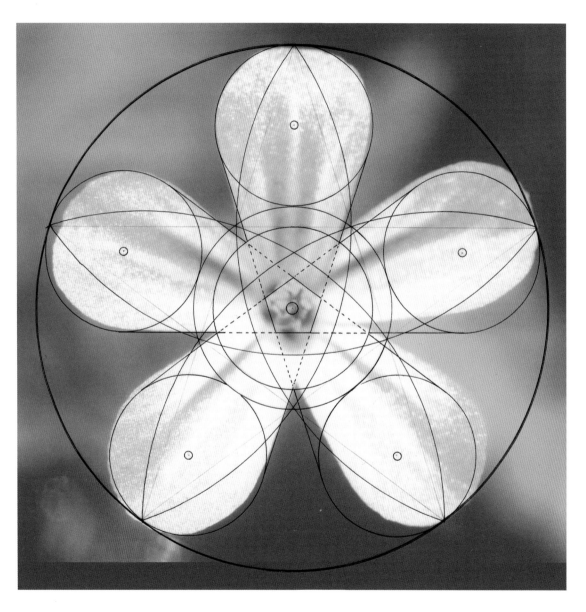

The Herb Robert has already been identified as one of my most favourite flowers.
Its very modesty and unusually powerful symmetry means
it stands out in any garden context although it is tiny.

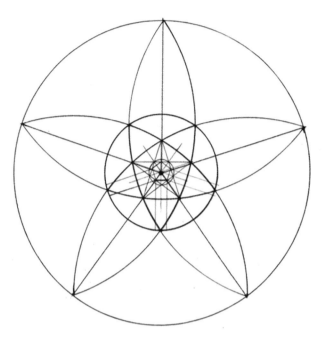

Another tiny favourite in my childhood is the Forget-Me-Not. The geometry of this remarkably proportioned small flower is startling in its conformity to pentagonal symmetry. It is clear that the centre of this flower is a decagon or ten-pointed white star. The parallel white extension on the larger image (following page) can be derived from the central star pentagon.

This wonderful little flower has individual features that charm
the eye yet it remains insistently fivefold. The inner colouring of
deeper red gives us a tenfold pattern.

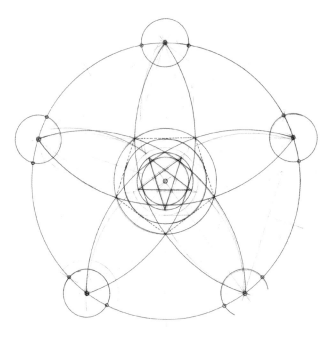

The geometry is based on the pentagonal geometrical flower. The circles on the
tips of the petals come from the centre. The tiny circles are the centres of the
petal curves of the neighbouring petal.

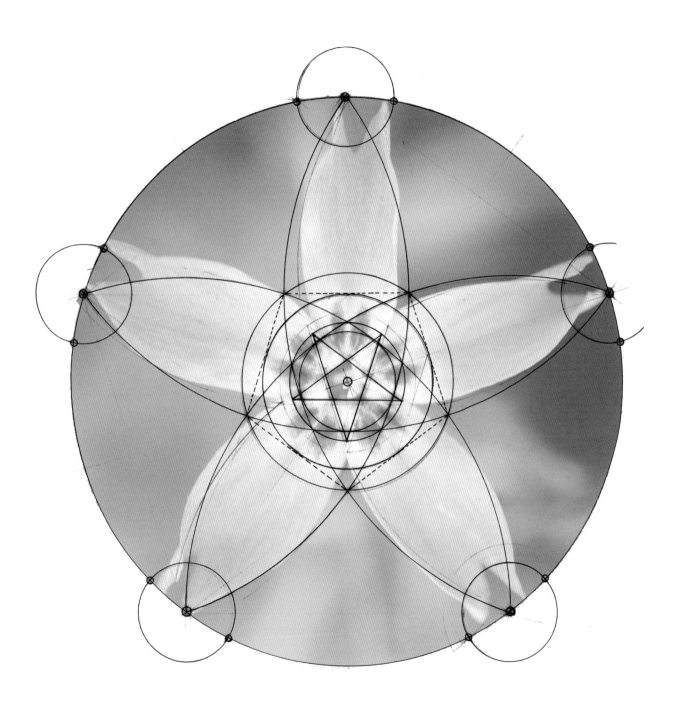

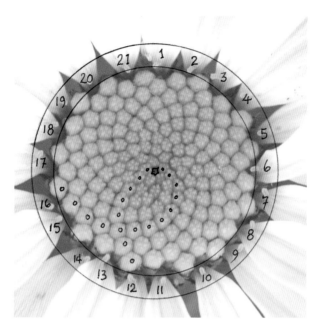

Here is another member of the Bellis family. In the centre we find an accurate golden number relationship of 13 florets spiralling in to the right and 8 florets spiralling into the left. 8, 13, 21 are in the golden sequence.

AS LINES OF
PROGRESSION

AS CIRCLES

AS POLYGONS
CLOSE FITTING

Sometimes analysis is clearer in one direction as in this Daisy face. The spirals demonstrate hexagonal close-fitting as well as a linear number spirals.

The proportion of this Daisy is pentagonal. To find the central area of yellow we
have placed four pentagonal stars to approach the twenty-one petals.

The symmetry of the pentagon holds many secrets and semi-hidden patterns
(seen here in a Geranium). It is by far the most popular of the symmetries
for the 'wild flowers', maybe because regular fives so aptly embody the golden
proportion. Here we see how the tenfold symmetry 'controls' the petal curves.

This modest yet powerful five-petalled flower in
the Pink family displays its colouring in such a way
as to reinforce its fiveness.

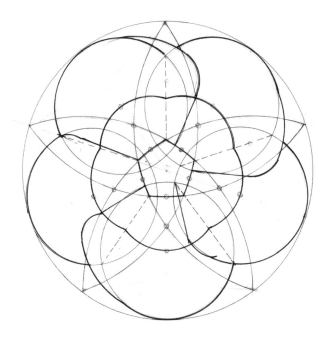

Close scrutiny by the geometer will reveal sources of
the petal and colouring curves.

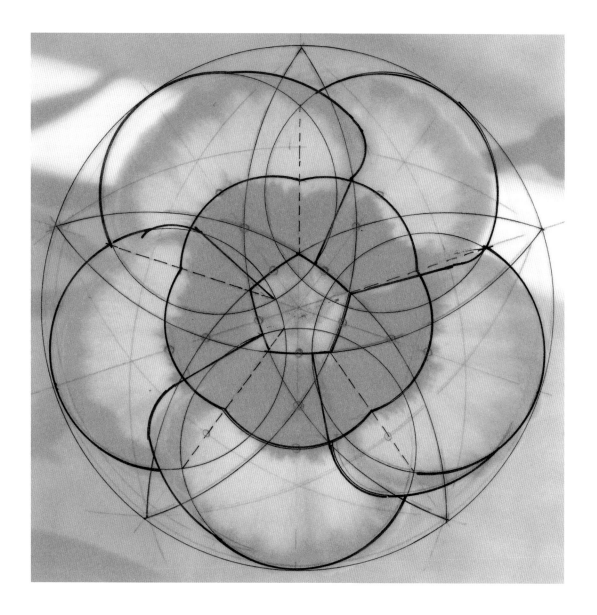

This flower, which is prolific in spring and summer in
my garden, is almost the classical representative of the
geometric 'five' flowers.

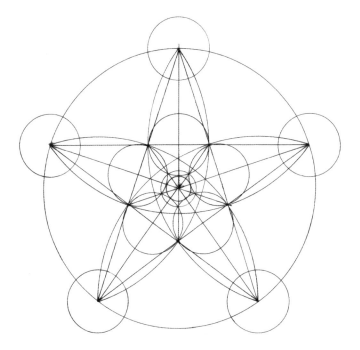

The five circles in the centre of the pattern are centred on the half-way points of
each diagonal of the largest star pentagon. They recur again in the petal tips.

The geometric analysis of this common small wild flower can only be offered at
an initial but important stage. A close scrutiny of the positions of the centres
that determine the curves of the petals repays the effort — yet also increases the
mystery. The ragged 'ends' to the petals we offer to others to solve.

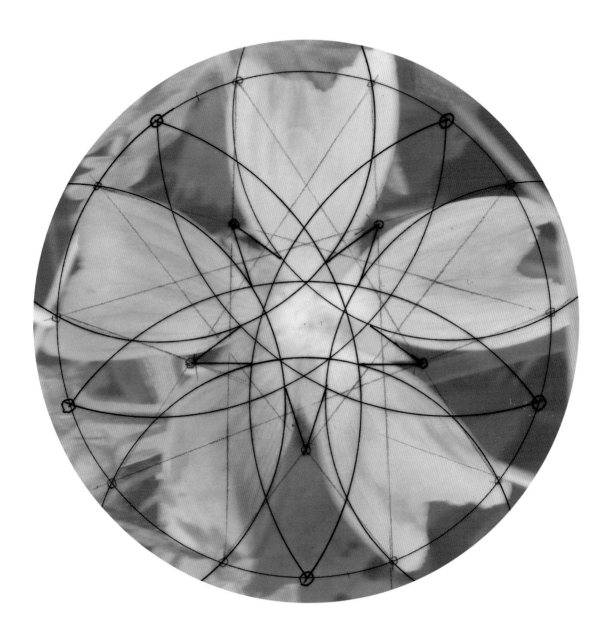

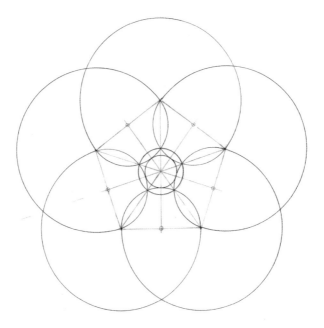

Those who have grown the Shrubby Mallow or *Hibiscus* in their garden will know of its dramatic effect, reaching well beyond human height. Here we offer an analysis of the central colour changes of these large fivefold flowers.

Careful scrutiny of this simple yet subtle geometry shows
how well it aligns with this beautiful little pink flower.

Here we show the fivefold geometry that offers a proportional solution to this
stage in the unfolding of a Rose bud. The simplicity is disarming!

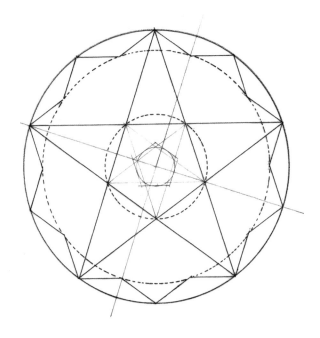

This colourful member of the Bellis family demonstrates a quite striking relationship with the proportions of the cross-section of the DNA helix.

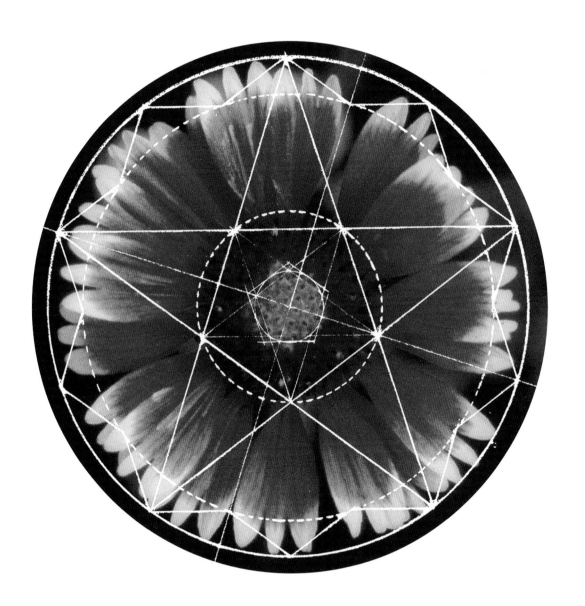

NOTE ON THE CONCEPT OF 'GENTLE EMPIRICISM' IN RELATION TO FLOWERS

The images that are most likely to come to mind for those who have studied the symmetry of flowers are those of sections cut through the body of a flower.

In this book we have chosen to follow the observations of Goethe and Rudolf Steiner. Both these observers of nature and natural science became alarmed at the taken-for-granted usual violence that is committed on the living world.

Botany for quite some centuries has used the unquestioned technique of killing life to analyse its material contents. This is the natural consequence of believing that everything has a material cause – even including intelligence itself. This meant that to 'kill' in the name of science was permissible. The stories of René Descartes' animal surgery are enough to raise the hind hairs of one's head.

So we choose not to encourage any further killing and dissecting of flowers. Rather, we have chosen to accept each flower as it presents itself to our eye (or camera). This decision is inspired by Goethe's 'Gentle Empiricism' which seems to us to be the only respectful and responsible approach.

We offer this drawing as a 'likely' archetypal Primula. This profile has much in common with the False Oxlip. There is much to be gleaned by the careful geometer from this study.

POTATO PLANT SECTION ANALYSIS

These sections through flowers, although beautiful in their symmetry,
represent an attitude of violence toward the living flower which we cannot approve of.
However, we present this example and our analysis — as an exception.

Flower sections are very inviting, but this is not the way the flower presents itself
to be seen. To obtain these, severing through the head of the flower is necessary.

This Wood Anemone demonstrates its own
version of utilizing the archetype of sixness.

These two colourful members of the
Primula family demonstrate clear sixness.

The radius of any circle divides its perimeter exactly into six. The resultant
compass-drawing shows the classical six-petalled geometric flower.

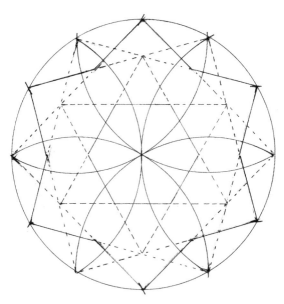

This geometry derives quite naturally from the primary sixness
which becomes a rhythmic twelveness.

The co-ordinated 'six-star' composed of right-angled 'points'
fits this Tulip colouring with surprising accuracy.

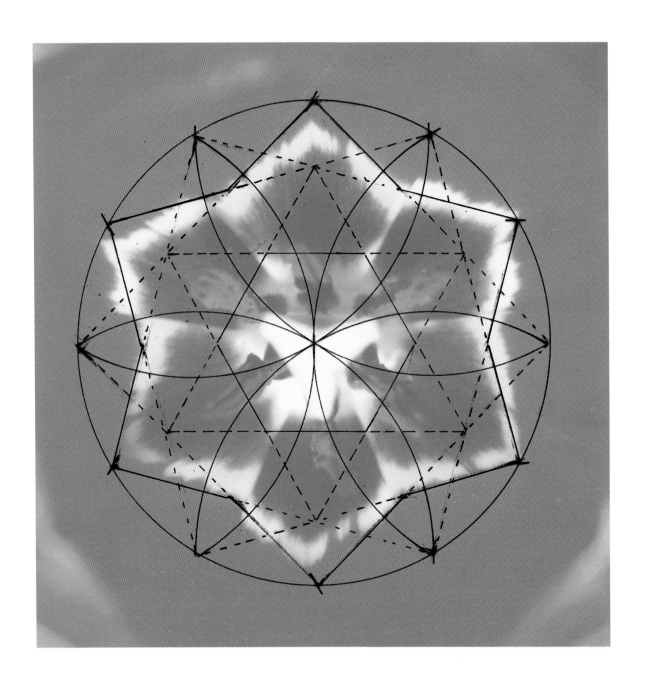

This is a fine sevenfold Clematis flower.

This Poppy seedhead demonstrates a dramatic sevenness which must surely have
aroused the enquiring human mind many times in its lifespan.

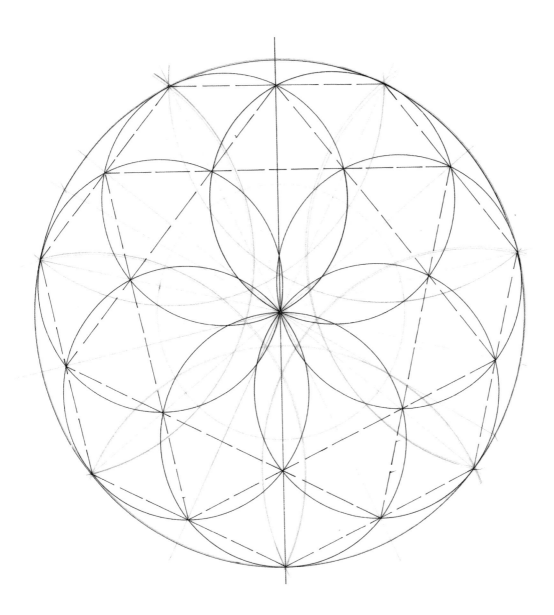

Here we have developed seven circles symmetrically arranged
with each a half-radius of the containing circle.

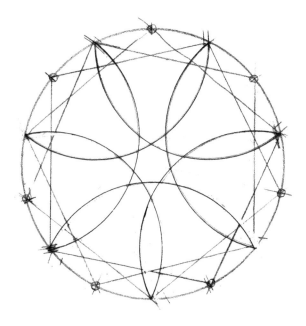

Here we find a rare sevenfold Tulip. We have constructed a conjectural sevenfold construction based on 14 equal divisions of the containing circle.

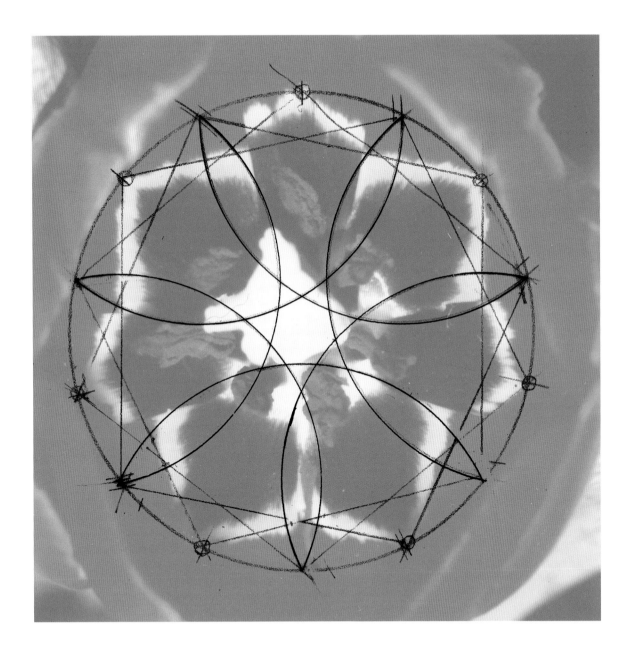

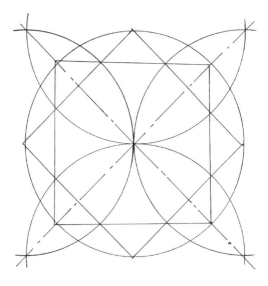

This exceptional eightfold Tulip pattern can be simply analysed by the
intersection of four equal arcs.

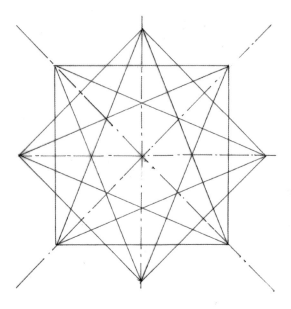

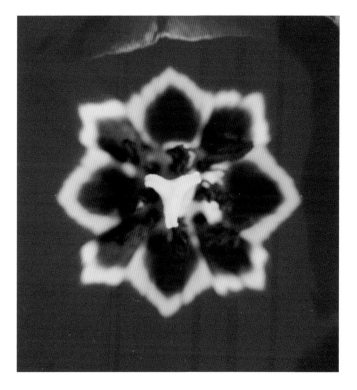

This Tulip being eightfold is again quite out of the ordinary and surprisingly
beautiful. We have given an interpretation in geometry
of the proportion of the inner colouring.

This version of eightness offers further interpretation and demonstrates that
the Tulip will explore sixfold, sevenfold and eightfold symmetry with its petals.
Maybe a reader will find a ninefold Tulip?

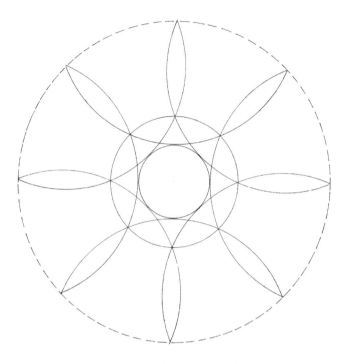

Here we show how this eightfold specimen of Sea Holly answers significantly to
the essential octagonal proportions. A favourite with bees — and no doubt with
good reason.

These four Cosmos flowers all demonstrate the
eightness chosen by each flower for its petals.

This Clematis flower has been shown before but once again it exhibits
a beautiful and strong eightness in both its petals and the spaces
between the petals. Both flowers are clear reminders to the enquiring
human eye of the pattern of eight.

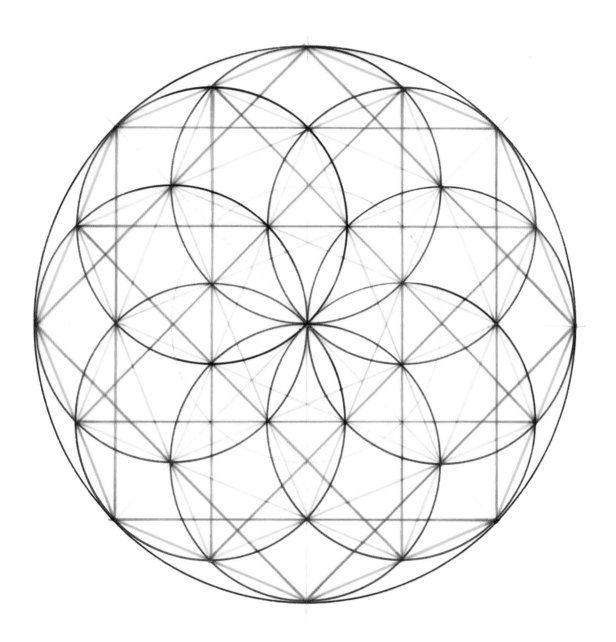

The half-radius circles within the whole are here shown in an eightness. At least
three kinds of geometric petals are implied.

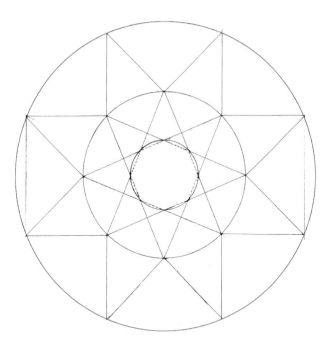

Osteospermum 'Whirligig', an exceptionally beautiful flower of the Bellis family (now familiar to readers), with its rare curling petals, demonstrates how well it conforms to octagonal (eightfold) proportions. This will echo the proportion of 1: $\sqrt{2}+1$.

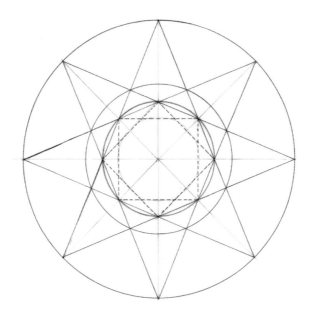

Looked at carefully, this Gazania flower shows quite clearly its basis of eightness
in the deeper red as well as in the pollen-laden yellow.
Geometrical analysis simply confirms what the sensitive eye can detect.

This yellow Celandine flower has a beautiful eightfold geometry like the sacred Lotus. Our analysis demonstrates that each yellow petal can be treated as an exact *vesica* although they overlap slightly with each other.

This *Passiflora* or Passion Flower decided to be on an eightfold pattern with its outer extensions. Our analysis shows that this symmetry was maintained by the radiant coloured fringes until the central fiveness of the anthers was reached.

It seems particularly appropriate that a Cosmos flower should be structured in an eightfold petal pattern. We show the route from fourness to eightness and have drawn a fourfold pair of geometric 'petals'. We have inscribed a square where these cross at right angles and proceeded to construct an eight-pointed star within a circle, crossing through the inner octagon which defines the central feature of this flower.

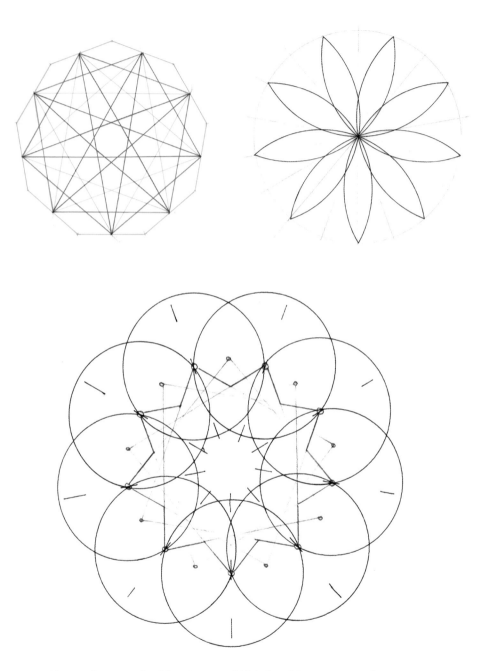

This is a fine example of the nineness which is found in some types of cactus.
For comparison, we also include the geometry of nineness in straight lines and
geometrical petals.

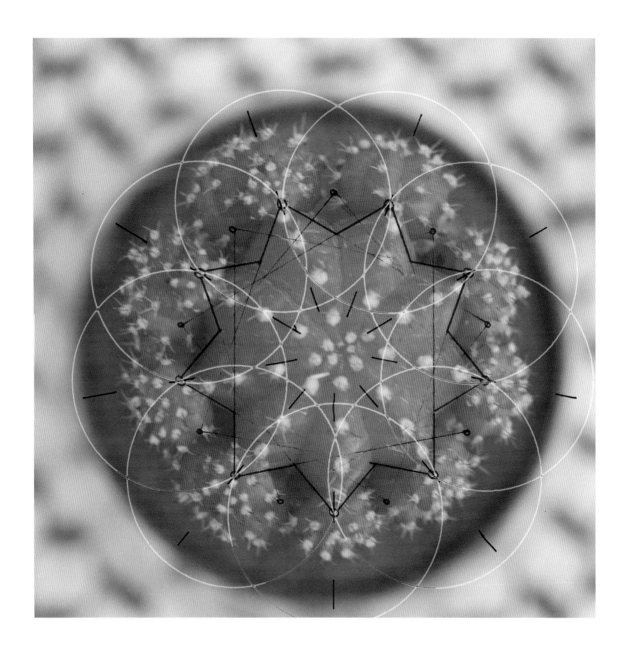

The *Passiflora* or Passion Flower most often has ten outer extensions.
The radiant colours that appear like fringes come next with a powerful fiveness
and threeness of central structures foremost.

The pentagonal geometry, supported by its intermediary tenfold intervals,
supplies very well the limits to the fringes of colour as well as
the central anthers and triple feature.

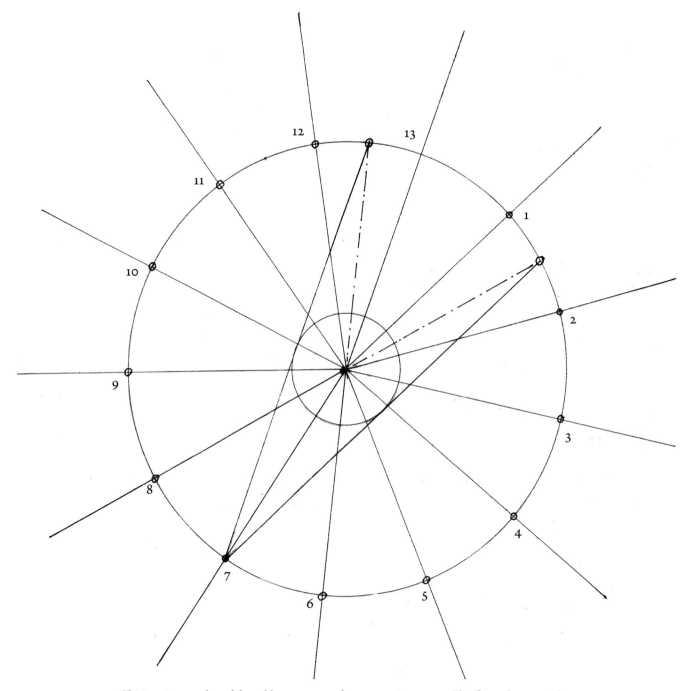

Thirteen is a member of the golden sequence of 1, 1, 2, 3, 5, 8, 13, 21 . . . This flower has 13 petals.
Here we have discovered the secret of the size of its central feature by dividing the circle into
2x13 divisions and thereby tangenting the central containing circle.
The points of tangent are on radii no. 10 and no. 4. This gives us the proportional
size of the circle containing the 'heart' of this flower.

The sacred Lotus offers its seedpod head as a cipher for awakening the objective consciousness in humans.
This is the prompt to the power of number. In this Lotus we are given a threeness surrounded by a nineness.

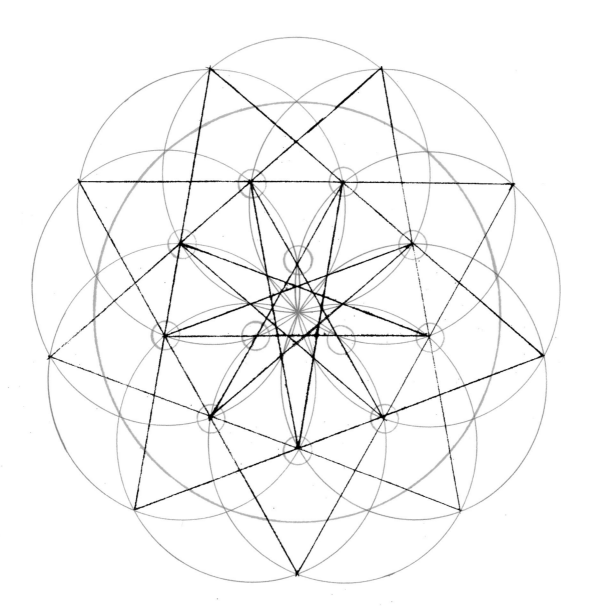

This analysis, based on straight lines and half-radius circles, offers an archetypal
interpretation of the seeds' positions in the seedpod head of the Lotus.

It was Albert Einstein who considered the human sense of wonder to be our highest gift.

The Lotus seedpod head having lost all its petals now poses a mathematical
question as to what 'number patterns' might intrinsically mean.
Is it the role of natural forms to open up in our minds the opportunity
of discovering objective ideas and truths?

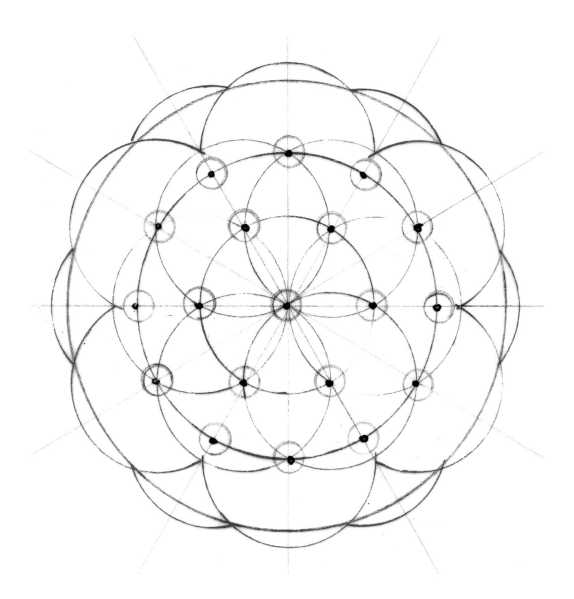

Here we offer a 'likely' geometry for the disposition of the seeds in the Lotus
pod on the previous page. The geometry is the natural outcome
of six circles all touching the central 'seventh' point.
Around these a natural twelvefold division occurs. The outer perimeter of twelve
arcs is struck from the given twelvefold pattern.

The Virgin Mary in the sixth position above Jesse in the Tree of Jesse window of Chartres
Cathedral. Jesus is above the crown of Mary. The genius of the glass masters of Chartres is
beautifully expressed here. Mary represents the stalk of the symbolic tree yet she also exhibits
her intimate relationship with the plant world. Note the sixfold flower just above her right
hand and the fivefold flower just above her left hand. The spirals, leaf forms and special curves
make this one of the great master-works of art in Europe.

FOUR

FLOWERS
OF GEOMETRY

It is impossible for matter to have any reality
without form, for it does not take on existence
unless it is dressed in form.

Solomon Ibn Gabirol

The Ancient of Days by William Blake. The image is of a
geometrizing figure holding a pair of open 'dividers'. Kathleen
Raine observed that this figure leans out of the sphere of eternity to
create a world in time and space.

Introduction to Flowers of Geometry

Geometry, as we would define it, is a universal and objective language. Before we enter into the meaning of the 'flowers of geometry' it is necessary to restate what our assumptions are. Geometry is the study of the order of space, one of the necessary five conditions of existence. This, in turn, brings us to two most fundamental tools for bringing the laws of geometry into experimental consciousness: these are the compasses (or dividers) and the straight edge or 'square'. They are likely to be the most ancient and revered of all scientific instruments. They embody actualities that can express 'absolutes' symbolically and directly.

One of the most revered spiritual practices of an advanced Zen monk is to make a circle in a single gesture with his ink brush.

These two tools guide the human hand into the realm of objective universality. This is in contrast to what is called 'freehand' drawing which is completely subject to the will and skill of whosoever's hand holds the pen or pencil. 'Freehand' work (human graphic arts such as Chinese and Japanese brush landscapes) is totally relevant to the psyche but it is of a different order from expressing and experiencing geometrical graphics. The word 'participation' was very popular with the later Platonic philosophers such as Proclus, Iamblichus and Plotinus.[45] This refers to practices — both theoretical and operative — where the human concerned *becomes the instrument* participating in an objective and philosophical proposition, not merely contemplating it or theorizing about it. This in turn implies that the human mind is capable of participating in a higher or superior intelligence. Some defined this as *noumenal* which is the realm of ideas that are 'permanently' true. The Pythagoreans, we assume, first posited that education should best be founded in the four unfoldings of number. First, pure number becomes arithmetic; second, number in space is geometry; third, number in time is considered to be music or harmony; fouth, number in space and time becomes astronomy, cosmology or spherics.

This demonstrates the degree to which an accomplished Zen circle is respected and used in mindfulness practices at a Zen monastery.

We advocate that all might 'participate' in the art/science of geometry in what might well be called, after Goethe, 'gentle empiricism'. We are under the complete guidance of the movement of the compasses as well as the rigorousness of following the discipline of the straight edge (or ruler). With geometry we 'participate' in the timeless truths of the products of 'straightness' and 'roundness'. As previously mentioned, Socrates affirmed that geometry was 'the art of the ever true'. To engage with such was considered of immense benefit to the soul because it is one of the most reliable 'reminders' of its origins and goal. The doctrine of *Anamnesis,* we repeat, was fundamental in Platonism. It means to 'recall' or 'remember' one's origins because therein lay the field of wisdom or the reuniting with the whole.

The compasses are governors of the circle and the circle is controlled by the still centre point. The circle is the most profound of all geometric shapes — being *all* and *nothing* — a good Zen koan!

The straight edge represents objective justice in that it offers *the most direct distance between two points* keeping an exact balance either side of straightness.

The human eye has an extraordinary capacity for recognizing straightness through its line of sight. The human eye has apparently been demonstrated to be able to judge accuracies up to thousands of an inch.

As we have already said, the circle of the compasses represents the 'heavenly' or intelligible realm symbolically. The straight edge or ruler represents the sensible or 'our' world of manifestation. The straight or orthogonal is measurable, while the transcendental nature of *pi* (π) of the circle is fundamentally *non*measurable or *arretos* according to Plato. (The translation of words like *arretos* to mean 'irrational' does not follow Plato's own usage. *Arretos* means 'beyond human explanation, ineffable' — not 'irrational' but rather '*trans*rational'.) Needless to say the 'reality' of geometry does not lie in drawn lines — which are merely symbols — but in the universally true relationships of conceptual space. The instruments of geometry merely follow the universal laws inherent in them. Drawn geometry is only an introduction to the deeper timeless meaning of order, which is cosmos itself.

The miracle of creaturely life is sight. We tend to be so familiar not only with our own eye in the looking glass but also with all the other important human eyes that we rely on during the day. Our eye, its iris and its pupil are all 'circular'. Plato pronounced sight as our most precious gift.

The Flowers of Geometry

For when I say beauty of form, I am trying to express not what most people would understand by the words … but I mean the straight line and the circle and the plane and solid figures formed from these by compasses, rulers and patterns of angles, perhaps you understand. For I assert the beauty of these is not relative … but they are always beautiful by nature.

Socrates in Plato's Philebus

In all human bodily actions there is either a conscious or a subconscious neural impulse which we can call the guiding idea or conceptual directive.

The Chinese discovered a minimal set of 'conditions', a threesome which they called the *Heavenly*, the *Earthly* and the *Human* as the essential 'triad' of consciousness. This means taking into account that which is *above* human understanding (the domain of pure unchanging principles) as the *Heavenly*, next that which is *within* human rational understanding as the basis of sensory experience and thought. Finally the *Earthly*, the behaviour of the physical and natural world, the Heavenly and Earthly being mediated and balanced by Human consciousness. It is this Grand Oriental Triad that the ancient Chinese believed held the purpose of life, namely mediation which *is* 'Being'.

Traditional systems have on occasions included a fifth or quintessence that is a central integrating factor of the four aspects of consciousness. (In his *Timaeus* dialogue Plato avoided any discussion of 'a certain fifth' which was the dodecahedron in his geometrical system, and it was left to Aristotle to offer it later as representing the 'Ether'.) The four elements are experienceable whereas this fifth is completely incorporeal and only available to the mind — which is probably why Plato only hinted at its existence and kept to the Pythagorean ethic (as tradition tells us) that it was esoteric and not to be discussed in the same frame as the other four elements or solids which are tangible.

The two triangles (called set squares in a technical draughting shop today) are the 30°, 60°, 90° triangle — claimed by Timaeus to be the 'most beautiful' in the universe — and the 45°, 45°, 90° triangle which is a half square. The validity of these two triangles selected

The fiveness of a flower's petals is usually very accurately fitted into a circle. This symmetry is a normal expectation from equal growth of each petal. The extra colouration on this small flower greatly enhances its beauty, and can be seen to be an extension of the intrinsic geometry.

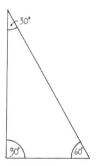

This apparently modest triangle, which you will find readily on sale in any draughtsman supply shop today, is in fact the one that Timaeus, Plato's spokesman in his cosmogony, said was the most beautiful in the whole of the created universe.

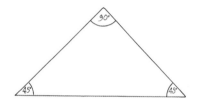

This 'set square' is also a common sight in any supply shop. How many of us are aware that Plato afforded this triangle cosmic significance in the crafting of our universe by the Divine Craftsman or Demiurgos.

by Plato in explaining the molecules of matter is reflected *in their continual practical use for draughtsmen right up to today.*

These two triangles carry two very significant and arithmetically mysterious qualities.

The first 30°, 60°, 90° triangle is half an equilateral triangle and has sides which measure 1, 2 and $\sqrt{3}$. This means that two of its sides are completely rational and *un*mysterious, but the third side is a key 'proportional' known as 'the square root of three'. This root dimension is incalculable by arithmetic or whole numbers, meaning that the solution will continue indefinitely and avoid completion — thus its mystery — *yet it can be drawn with precise accuracy geometrically.* As far as the human hand need go the points that the line terminates at are exact, but measuring this line arithmetically in whole numbers cannot be achieved. This triangle Timaeus (Plato) allocates to the 'construction' of the molecules of 'fire', 'water' and 'air' and declares it the most fair or beautiful of all triangles in the fabrication or ordering of the universe by the Divine Craftsman, the Demiurgos.

The second 'set square' or triangle of 45°, 45°, 90° has two rational sides — say measuring 1 and 1 — and the third side as mysterious as the first triangle, being 'the square root of two'. The *square root* of two, meaning that number when multiplied by itself becomes *two*. However this is also non-resolvable by arithmetic. Thus both numbers $\sqrt{2}$ and $\sqrt{3}$ can be called *trans-rational* or beyond rationality (as is intuition). Some historians have held that these numbers were 'threatening' to the Pythagoreans, for instance, who held Number in the very highest regard. However the truth is more likely to be that these dimensions were held as *arretos* and esoteric. They certainly knew of them. For those who want 'irrationality' to be the weakness of number they may have it, but it does not dismiss the *precision* and *significance* of a geometric expression of the same truth. It is far more likely that the trans-rational numbers were indicators of a 'different reality' for the Pythagoreans. Each may choose.

Let us return to the practice of Geometry and why we have called this chapter the 'flowers of geometry'. Geometry is simultaneously an art and a science. It is not valuable in the opinion of the author to separate them. The *art* is to bring them into awareness universally and outwardly, the *science* is the clear and precise nature of their principles, straightness, roundness and angularity.

A principle we must repeat is that there is a necessary connection and interrelation between the four subjects (later called the *quadrivium*) of the classical educational curriculum. That is between Arithmetic, Geometry, Harmony or Music and Astronomy or Cosmology.

This means that intrinsically they all *complement* each other in their own unique language. (In short four aspects of a unity.) Hence when we read of a cosmogony, that is a birth or coming into being of a universe, in any civilization's traditions, it will inevitably have a geometric expression, implied or actual.

In this instance we will call these 'cosmogonic flowerings', and we will take one of the ancient Chinese versions as well as the Biblical and Koranic. The first example will demonstrate the fiveness that predominates in the ancient Chinese world picture. The second will demonstrate the sixness that prefigures in the Abrahamic faiths. Both are examples of how the language of geometry is an adequate symbolic language for expressing cosmogonies.

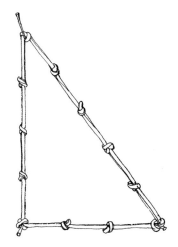

This triangle, another key scientific discovery, is composed from a cord with twelve knots in it. It is used on building sites today as it was thousands of years ago. Pythagoras's theorem is easily recognized in its form, as five squared is equal to three squared plus four squared (i.e. $25 = 16 + 9$). This form of geometry was believed to have been used by the ancient Egyptians who called it 'rope stretching'.

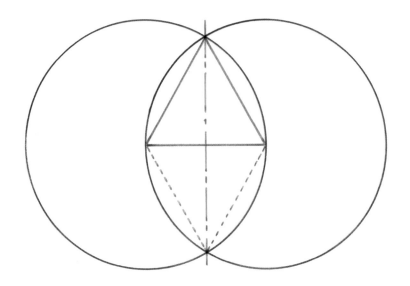

EUCLID'S FIRST
PROPOSITION DIAGONAL

This shows the simple importance of Euclid's first proposition to make an equilateral triangle. Euclid does not however point out that the vertical dashed line cannot be determined in whole numbers — as it is the square root of 3.

The compasses or dividers are humanity's oldest and most reliable scientific
instrument. Objective knowledge was born for us when the compasses
were discovered. Composed as it is of two 'legs' and a single hinging top,
it is completely in harmony with the human hand.

Geometry Begins with a Circle

Indian geometry after the Sulbasutra period grew up for the sake of the circle, the celestial circle.

Dr T. A. Saraswati Amm

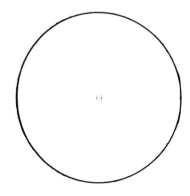

It is primary and inevitable that any geometric construction, particularly a cosmogonic one, will begin with a circle as it is the primary figure of the 'all containing'. The first exercise is visualize a circle, then place the compass point in the best position (centrally) on the surface. A little pressure is recommended for stability. Then 'turn' the compass from its apex. When a comfortable movement is achieved, lower the drawing point on to the surface and draw a circle.

What you will have drawn is the most elemental and profound geometrical figure. It is a line in which every part is equally distant from the centre that you first established with your compass point. There is always more to such a first drawing, depth upon depth, but for now we can consider we have a profound image of 'oneness': a single circle, a symbolic *universe*. The paradox is in the profound simplicity. Remember the 360 degrees of a circle; each degree represents a day in an archetypal year, 360 of the actual 365 days. This symbolism was established in ancient Egyptian times and probably even before.

In the first instance we are going to follow the ancient Chinese cosmogony (birth of the universe) and arrive at the geometric 'flower' of fiveness. This accords with the Chinese 'pentagonal' universe. It will however inevitably follow through the number sequence of 'twoness', 'threeness' and 'fourness' on its way to the final goal of five. The ancient sages all sought the inner significance to each move in these stages of unfoldment. Geometry as an art was considered a form of participation with the Tao or divine intellect or timeless principles. 'Coming into Being' was very seriously considered.

'Tao like water always takes the lowest position'. (*Tao Te Ching*) Humbleness being cardinal to profundity.

Although the universal timeless principles 'contain' all and permeate all they are intrinsically humble and always there, like water, essential to life.

The circle is the archetype of archetypes in the geometric sphere. It embraces all other shapes and is thereby all-inclusive. It is the shape of the two most important bodies of light in the heavens: the sun or full moon. As Proclus in his commentary on Euclid says of the circle, it 'regulates all things for us down to the humblest bounty, as it dispenses beauty, homogeneity, shapeliness and perfection'. It is the shape of our eyes and as a sphere is the shape of most fruit.

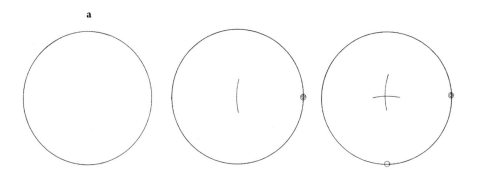

FINDING THE CENTRE OF A GIVEN CIRCLE

Finding the centre of a given circle '**a**'.

First one places one's compass point on the right-hand side of the circle on the point given with a small circle.

Strike a small arc across the centre point with your compass set at the same radius as the given circle.

Next place your compass point with the same radius on the bottom of your given circle.

This is marked with a similar small circle.

Draw a second arc crossing the first. This will give you the centre of circle '**a**'.

'Quickness' is never necessary.

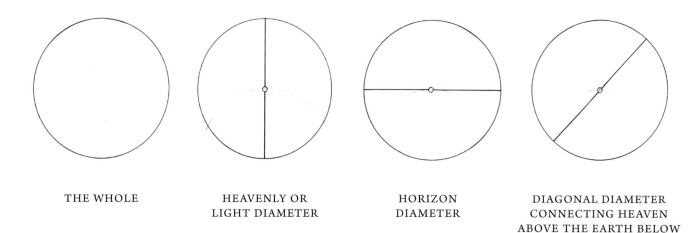

| THE WHOLE | HEAVENLY OR LIGHT DIAMETER | HORIZON DIAMETER | DIAGONAL DIAMETER CONNECTING HEAVEN ABOVE THE EARTH BELOW |

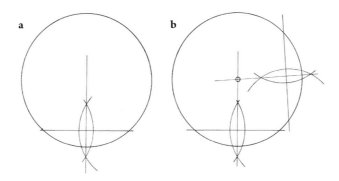

FINDING A CIRCLE'S CENTRE BY TWO ARCS

A way of finding a 'lost' centre in a given circle, particularly useful if one no longer has the radius.

a. Take any segment smaller than a diameter, halve it with a compass, draw long centre line past the obvious centre.

b. Draw second segment, halve it similarly with two arcs of compass. Extend central axis until it has crossed the first central axis. This will be the circle's centre.

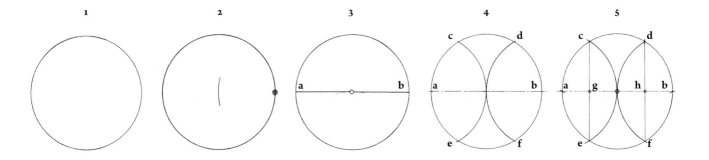

THE FIRST EXPLORATIONS OF THE CIRCLE FINALLY FINDING THE
TWO INNER CENTRES OF TWO SMALLER CONTAINED CIRCLES

1. The first simple circle.
2. Marking the centre.
3. First horizontal diameter **a, b**.
4. Next draw two radius arcs using **a** and **b** as centres which cross the outer circle at **c, d, e,** and **f**.
5. Connect **c** with **e** and **d** with **f**. These cross the horizontal diameter (**a, b**) at **g** and **h**.
   Thus **g** and **h** become the centres of the two inner circles [**e** and **f**] on page 301.

If we follow the captions to the drawings carefully we will eventually have our archetypal fivefold flower which we will choose to call the *'flower of life'*. The reason being that it embodies the golden proportional series so predominant in all living creatures. This is also a reminder that the 'world' or universe was considered to be a 'living' creature with a soul by the ancients. In the ancient Chinese cosmology we have the creation of the four seasons in the 'fourness' division we created. This ancient world picture then talks of the five 'Elements'. These are *fire, wood, metal, water* and *earth* and can be easily written in single words but have depths upon depths of meaning, due to their intrinsically subtle nature, that interpenetrate so many areas of human life. In the Chinese system these five were called the 'Elements' or the 'Agents' of materiality. The ancient Chinese art of acupuncture is also based on the five 'Agents'. The Chinese treated the human body as no less of a microcosm than the other ancient civilizations. Some commentators here used the word 'gold' for the 'metal' element of this system.

However this is not the appropriate place to go into further detail as there are many excellent books and even more excellent practitioners of this art for those who are interested to find out more.[46] Wing-tsit Chan is our recommended author and translator on Chinese philosophical matters.

At this point we would like once again to draw attention to the five-foldness of the great majority of wild flowers as well as the shared fiveness of hands and paws of so many animals. The integral symbolism of five-ness has a profound richness. For some this fiveness on both hands is the basis of counting up to ten and then repeating the sequence from one to nine again. And so on in pairs of digits. However there are other ways of viewing this. What is important is to avoid saying we have five fingers on each hand, we only have *four* plus our thumb. Yet both our hands and feet have an intrinsic fiveness in their overall extensions.

The next cosmogonic pattern we propose to look at, from the geometric view point, is the Creation story according to the tenets of the three Abrahamic faiths, Judaism, Christianity and Islam, based as it is on an active sixness and a passive sevenness. It is surely a mark of the universality of number that there is unanimity between these three 'faiths' on the matter of the creation taking place in six 'days' or intervals.

A seventh arises as either a central 'throne' or a 'day' of rest. In this instance it is also relevant to repeat that the most ancient of the Chinese books, the *Book of Changes* (*I Ching*), also begins its mathematical cosmic structure with what is called a 'hexagram' of six lines. This *Ch'ien* pattern of an equal sixness is named 'the Creative' and is the *first* of the

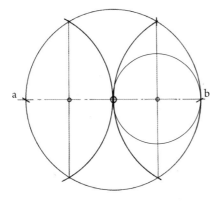

THE PROCEDURE TO FIVENESS FROM ONE CIRCLE

Firstly we find the quarter divisions of the diameter **a b** by drawing two *vesica* arcs and joining the tops to the bases of these. The vertical lines cut the diameter **a b** into four equal divisions, which includes the centre of the circle. We have marked these with tiny circles which will become **e** and **f**.

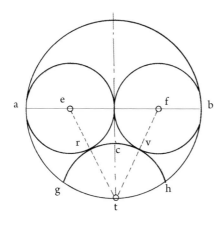

When we have found the centres **e** and **f** on the line **a-b** we draw two circles that touch both the centre of the circle and the circumference at **a** and **b**.

Now we have drawn two circles within our Tao. These circles have exactly half the radius of the larger containing circle thus can become the basis of the Yang and the Yin. So from twoness, the line, we now have a 'threeness' of circles — two inner and one outer embracing circle.

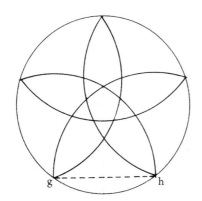

When we have found the radius of arc **c** centred on **t** which touches exactly points **r** and **v**, we cut our containing circle at **g** and **h**. This distance **g** to **h** is the exact fivefold division of our first larger circle. It is from **t** that we find exact balance between the Yin and the Yang circles that we can obtain our eventual fiveness which will become the geometric flower of five or as we have chosen to call it the 'flower of life'.

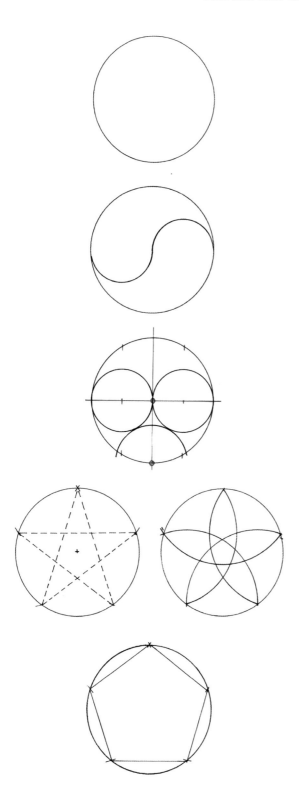

A CHINESE PERSPECTIVE

The geometric evolution of the ancient Chinese cosmogony. From the circle (*Tao*) to the Yin Yang, requires two circles of half radius within the larger circle.

From the two smaller circles we find the lowest point on the largest circle to strike an arc that exactly tangents the Yang and Yin circles. This arc as we found in p. 301, gives us the exact division of fiveness within the largest original circle. This fiveness represents what the ancient Chinese called the five 'elemental forces' or; water, fire, wood, metal and earth.

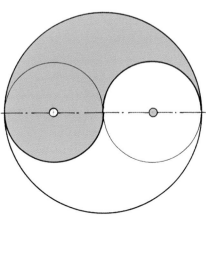

The division of the circle into two 'dynamic' sections typifies the ancient Chinese wisdom. One of these smaller circles is the basis of the Yin symbol, the other smaller circle is the basis of the Yang symbol. These two forces, 'passivity', and 'activity' are in constant interaction throughout the whole of the universe (Tao). The centres of the two circles require emphasizing. The Yin circle has a 'full' centre point while the Yang circle has an 'empty' centre point (very small circle); these are the acknowledgements of the opposites within each 'force'.

The most ancient objective system of seeking the possible combinations of positive Yang and negative Yin activity was devised at least three thousand years ago in China. Called the *I Ching* (*Book of Changes*), it is based on sixty-four hexagrams. Here is number one, the *Ch'ien* hexagram (drawn by a contemporary master calligrapher) for the creative principle, made up of six Yang lines.

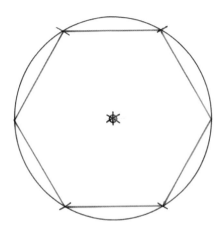

The sixness of a circle is precise. When the points are connected we have a hexagon. The centre point, at rest, represents the seventh.

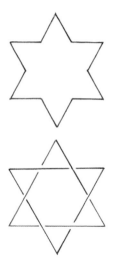

a. The sixfold star can be drawn as twelve continuous folded lines returning to the starting point.

b. Or it can be drawn as two totally balanced interwoven equilateral triangles.

Geometry as utility: the ancient skill of weaving was acquired when we first needed to create surfaces or containers for domestic use.

sixty-four possible combinations of Yin-and-Yang-lined trigrams. *Ch'ien* has six full Yang lines. This can be represented by Yarrow stalks and if rearranged geometrically can be set to make the image of a regular hexagon or a star hexagram by crossing over. It is also possible to construct the primary platonic figure — the tetrahedron — with these same six stalks as the edges of the solid figure. Each of these coincidental facts point to the universality of sixness, particularly with reference to creation.

So we will proceed to look at this geometry as a flower of six! This means a construction in the sense of creating similar circles that move outside and beyond the first circle. Here we can see geometrical interpretations that not only become profound metaphors for the scriptural creation stories but demonstrate that *the perfection of a single circle lies in its accommodating six of itself around its perimeter.* We can call these 'days' inasmuch as each of the six establishes a reflection of the originator exactly and symmetrically in six equal directions.

The sixfold flower at the centre we choose to call the 'flower of creation' by adoption and appropriateness. It has a curious reflection of the 'perfection' of sixness in the arithmetical behaviour of 1+2+3=6 and 1x2x3=6. The geometry exactly reconciles the arithmetically irresolvable *pi* (π) which the moderns call 'transcendental'. The perfection of the circle is thus geometrically matched (inherently) by the perfection of the sixness of its perimeter. Again the geometric construction avoids any of the non-conclusiveness of *pi* (π).

This sixfold flower has the important function of reflecting the 'Lily of the Field' so praised and exalted by Jesus when comparing it with 'Solomon in all his finery'.[47] The Lilies have sixfold symmetry to their petals — even if they often express this in two threenesses.

This first excursion into creation myths leads us back to a consideration of Unity or the One which is after all the intrinsic nature of any universe.

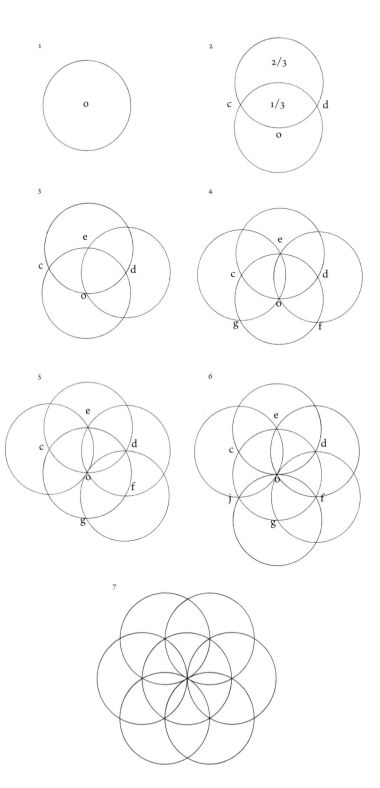

THE PROCESS OF DRAWING THE
FLOWER OF CREATION OR SIX
CIRCLES FOCUSED ON A CENTRAL
SEVENTH.

1. We naturally return to our originating circle.
The symbol of completeness and timelessness
— even the creative void.

The central point takes up the cardinal
position yet is not necessarily manifest or 'seen'.

2. By keeping our compasses (with which we
have just drawn our primary and primordial
circle) at the same radius, we now proceed to
find the point which had established the centre.
We choose the point at the top of our first circle.
Next we resume drawing a second circle of the
same radius.

The perimeter of this second circle will be
2/3rds outside (above) our original circle and 1/3
within. It will also pass through the centre point
of the original circle if drawn accurately.

This naturally demonstrates the cardinal
importance of this first construction. It is the
basis of the first proposition of Euclid's elements
of geometry, a proposition that has obvious
philosophical meaning for the discerning.

3. The two places where this second circle passes
through the perimeter of the first circle we will
call **c** and **d**. The next procedure is to place our
compass point onto the position entitled **d** and
proceed to draw a third circle.

4. This will cut our original circle at the top
which we call point **e** passing through the
centre again at **o** and cutting the original circle's
perimeter at **f**.

5. We now draw a fourth circle from this centre,
f, with the same radius. This will connect point **d**
to the centre **o** and proceed to cut the perimeter
at **g**.

6. We continue to proceed in this manner by
using point **g** as a centre, then point **j** as a centre
until we get back to the point **e** at the top. This
will complete the whole sixfold procedure.

Oneness
or the Quality of One Itself

. . . from One that does not emanate but is the Principle of Emanation of Life, of Intellect, and of the Universe.

Plotinus, Enneads III 8

The incarnate Word is with us,
is still speaking, is present
always, yet leaves no sign
but everything that is.

Wendell Berry

The reflective bubble floats in the garden. A familar delight to all mature children as well as children. A single bubble naturally becomes a sphere by using the minimal surface to embrace the greatest amount of air. Water and air express integrally and geometrically the archetype of archetypes.

Unity is at once the simplest concept of all mathematics yet the most embracing, and like zero the most mysterious and elusive. It seems that there is a ladder for our awareness to climb which starts from one thing or one object as one part of our environment and can lead us both down into multiplicity or up towards unity as an integral understanding.

We can take a step up when we realize that this singularity is common to the experience of those around us; they also see or feel the same one thing, albeit from slightly different personal viewpoints.

This becomes a further rung to the ladder when we find that a whole culture (even many centuries old) takes the same view of this oneness — remembering that however large a number may be it still only sums to *one* number.

This can eventually bring us to the topmost rung where we can become conscious of the experiences of 'oneness' as the sum and totality. A oneness that *includes* both the object and ourselves as the subject. Oneness then becomes a profound experience and has been called atonement which is at-one-ment. We not only count one but are stimulated into the intuition that this oneness is a *quality* which can be experienced and is not just a matter of counting. Thus the ladder of awareness or consciousness rises before us as a consequence of our curiosity and natural wonderment. Number is obviously qualitative as well as quantitative, symbolic as well as factual. Qualitative on the ascending, quantitative on the descending, one might say.

Surely there is one Tulip in front of me yet there is a 'unity' or self-similarity between all Tulip-faces that we also recognize. Sixness is powerfully present in the Tulip as petals and as a hexagon — yet this is not totally invariable.

This identification of an abstract quality that connects all Tulips is a major step in our becoming aware of how we learn about principles which are 'above' and 'beyond' mere multiple experiences. This may seem trivial at one level but not at another. Sameness and difference are integral in being equally present and complementary. It is the recognition of *repetition* in our lives that offers us security. We rely on the sun rising each morning and we come to recognize not only the beauty and security of all similar daybreaks in our lives, but we realize that the whole of our existence is based on the repetitive rhythms of cyclic time. Yet we also recognize that although similar in essence, each day is different from each other day. Unity is the evident whole of life, duality its pulsating expression.

Oneness is not uniformity, yet one Tulip is also all Tulips — archetypically.

It is a curious fact that the Vedic cosmology implies the same necessity of threeness. The three are called the *Gunas: Tamas, Rajas* and *Sattva* which can be translated as inertia, energy and light. When we come across such simplicity we often let it slip by as if it were trivial, or so self-evident that it doesn't require consideration. However as we ascend our ladder of consciousness we become more and more aware of the profound simplicities contained in principles that apply universally. It is similar to finding the trunk of a tree having spent so much time amongst the twigs and branches. It is a relevant question to ask: is a science that studies only differences as valuable as that which encourages the study of unity, or even studies wholeness itself?

Unity, oneness, singularity, unicity, all refer to aspects of all-embracing essence of our experience and our *uni*verse. That is the very integrity which we call 'One'. There is also a beautiful simplicity in this 'single verse' we call universe — as if it is a poem in which we live, love and have our being. It is the value of poets and their privilege to express this.

> There is no end, no ending — steps of a dance, petals
> of flowers
> Phrases of music, rays of the sun, the hours
> Succeed each other, and the perfect sphere
> Turns in our hearts the past and future, near and far,
> Our single soul, atom, and universe.
>
> Kathleen Raine, from *The Sphere*

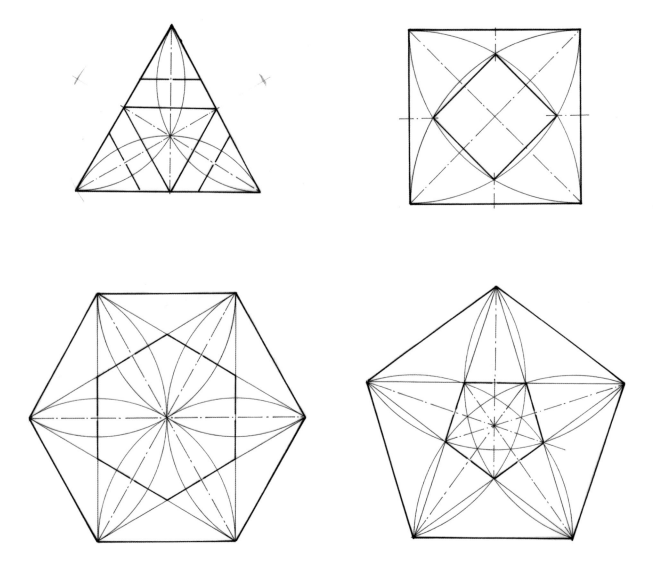

The triangle from threeness; the square from fourness; the pentagon from fiveness;
the hexagon from sixness. These simple shapes all combine specific geometric 'flowers'
generated by the arcs within.

From the previous illustration, we can see further how detailed
proportional growth or diminution can be created from each of the
respective symmetries of each circle.

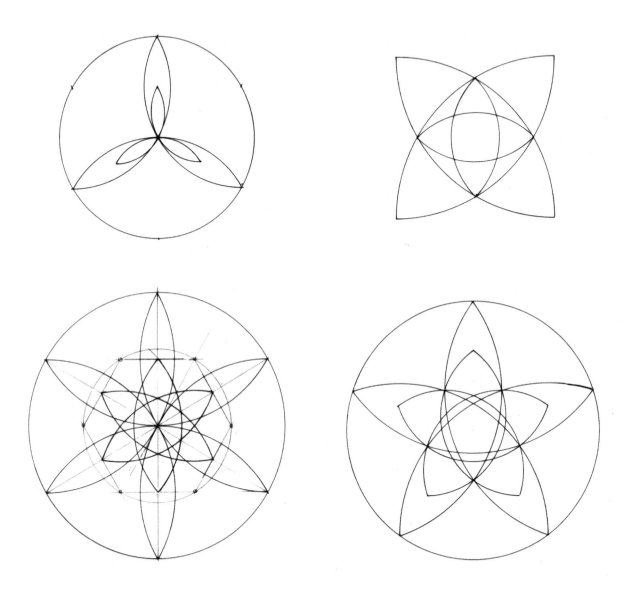

Here we show proportional elaborations of our four basic shapes
or symmetries as curvilinear 'flower petals'.

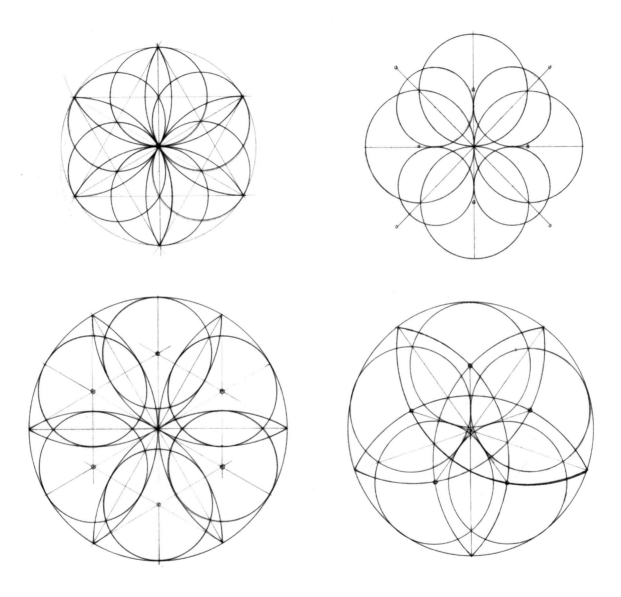

Now we elaborate further with our chosen symmetries which hint
at petal patterns and flower arrangements.

TWONESS

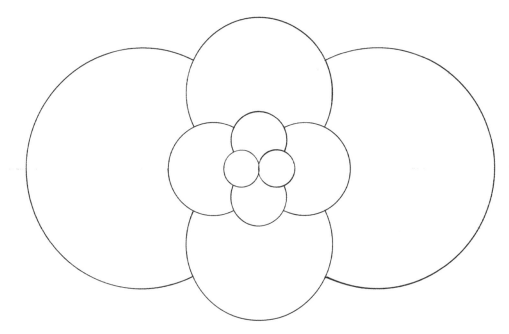

From the simple single circle we arrive at duality or complementarity,
with the proportional relations between pairs of circles.

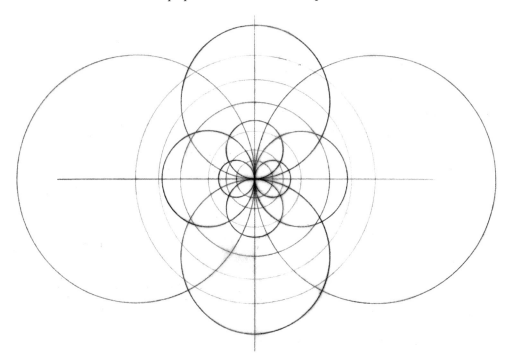

This drawing elaborates on the previous and shows inner structure.
A progression of 180° phyllotaxis.

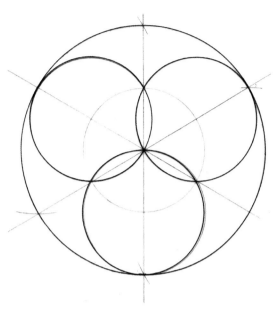

Next we will explore the patterns of circles within a circle or parts within a
whole. Here we take a given large circle and draw three further
circles of half radius within.

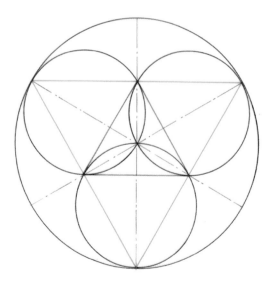

Together we see how the straight lines demonstrate the alignment
of the points of crossover of the circles.

THREENESS

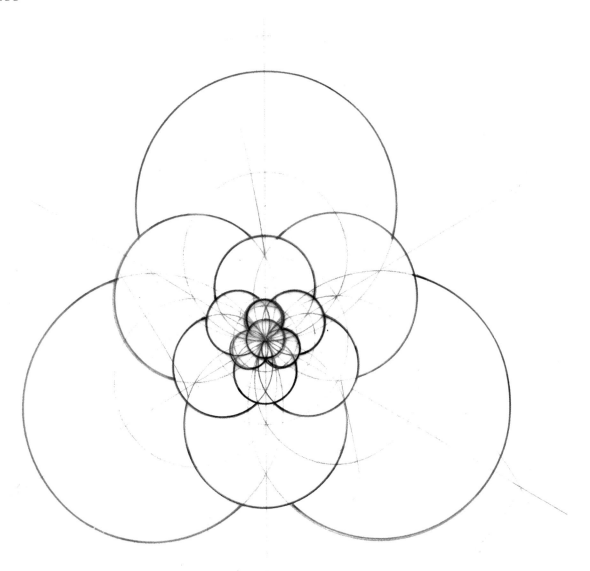

In this drawing we have shown proportional growth or diminution in
half radii of circles in threes.

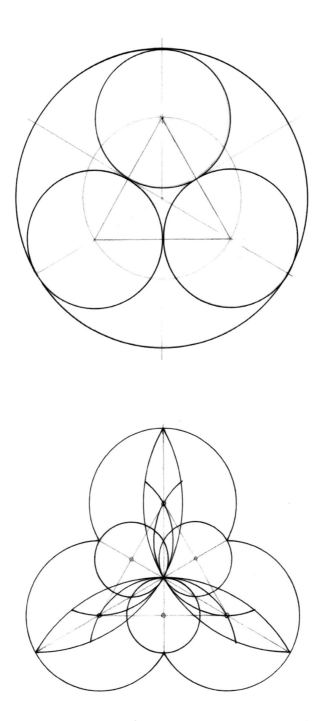

Here we have explored the inner geometry of threeness as a geometric flower.

FOURNESS

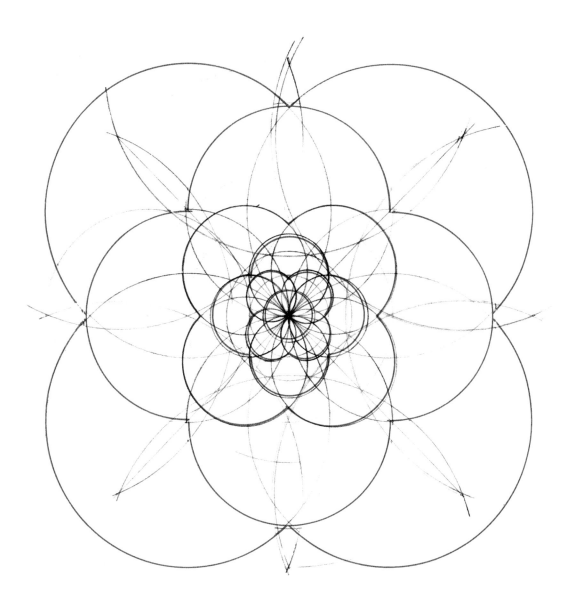

In this drawing we explore the fourness of the
proportional growth of petals. This pattern is based on
four symmetrical circles overlapping.

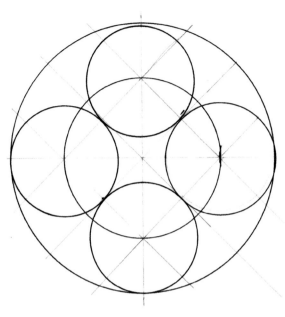

This pattern shows four symmetrical circles just touching.

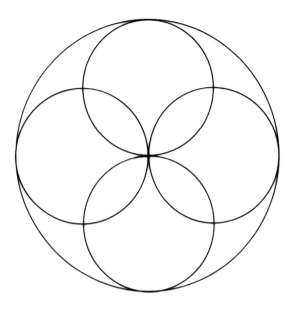

Here four circles of half radius within the parent circle
show the basis of overlapping (petals).

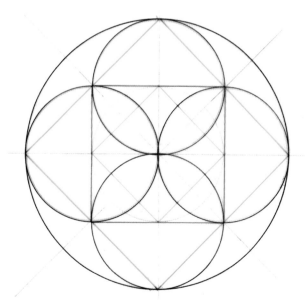

Here we show the geometry of touching outer points
and circle crossings in straight lines.

FIVENESS

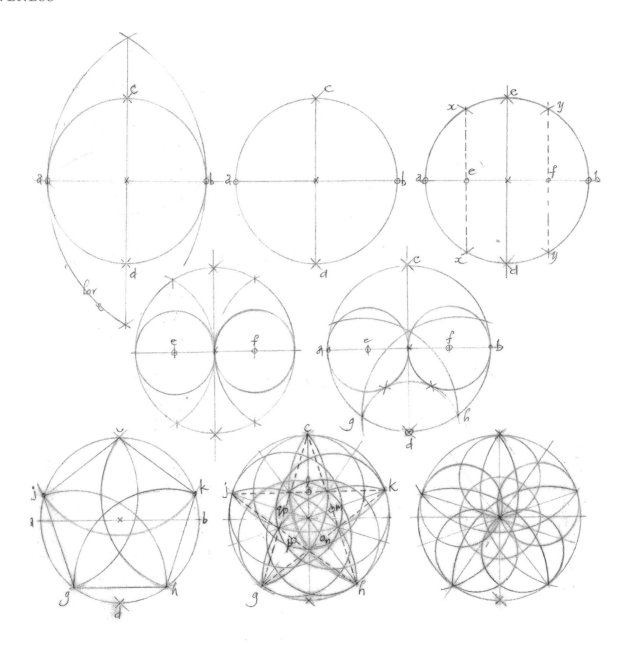

This page is a geometric study of the evolution
of creating a fiveness in symmetry.

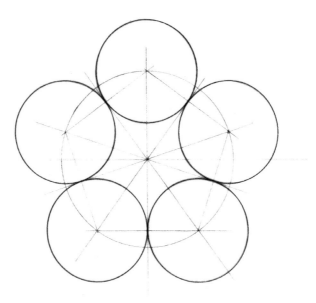

In this drawing we have placed five smaller circles symmetrically
so that they just touch. This means that another proportionally
smaller circle can also fit in the centre.

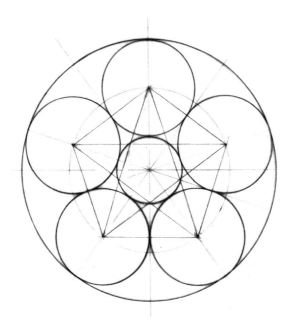

By drawing the pentacle between the five smaller circles we find
the middle circle exactly cuts the pentacle crossover places.
A proportion which occurs universally in our world,
from mean planetary orbits to the DNA-molecule section.

FIVENESS

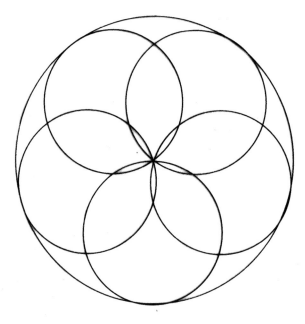

In this drawing we have now introduced five smaller circles into the parent larger circle. They are arranged symmetrically and for the first time we get two crossing-over levels between them, and thereby three petal types.

Together with this we also introduce the straight alignments which embrace both orbits of crossovers. This co-ordination is profound yet so easily seen when drawn carefully.

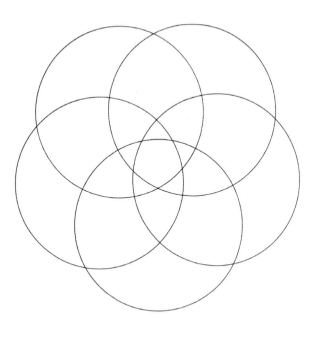

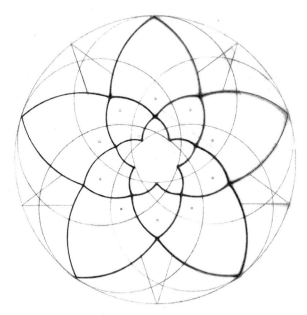

Speculative exploration of flowers of geometry in fivenesses.

FIVENESS

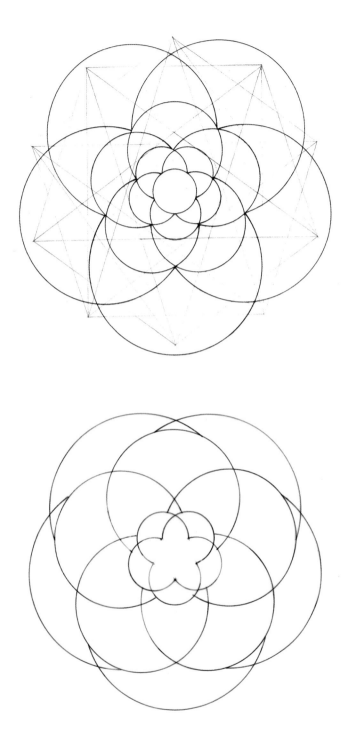

Speculative exploration of flowers of geometry in fivenesses.

FIVENESS

Geometric explorations of geometric arcs of fiveness, proportionally diminishing.

FIVENESS

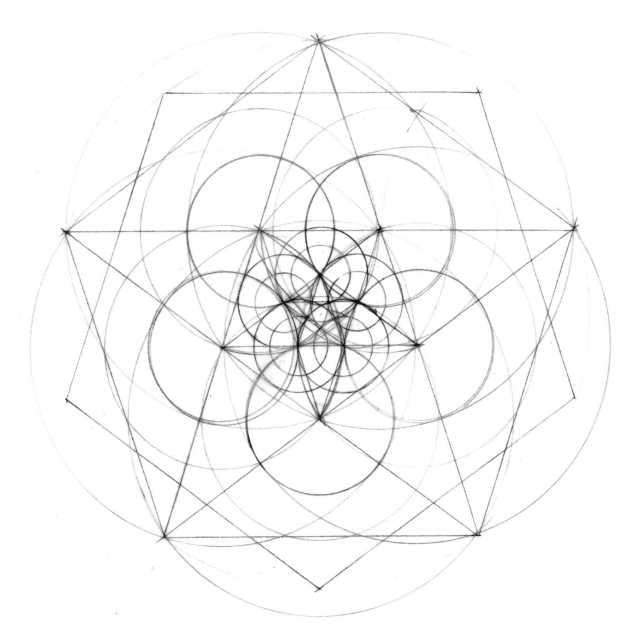

This drawing explores the interconnectedness of straight lines and
circles in geometric flowerings. The 'underdrawing' has been left visible for the
more particular reader or researcher.

This is another approach to fivefold symmetry in geometric flowering,
using only parts of circles intersecting.

SIXNESS

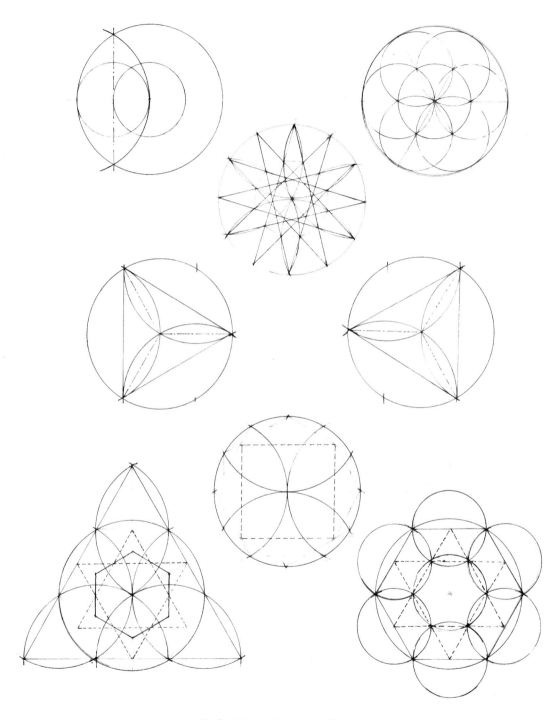

Explorations into aspects of sixness.

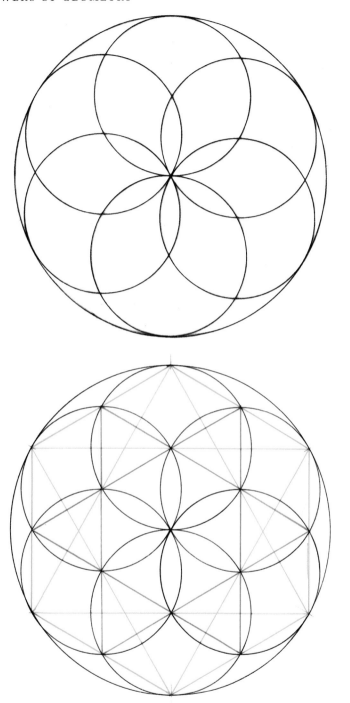

Here we follow our plan and demonstrate what six circles within a parent circle
look like. The smaller circles are half the radius of the larger. Also we introduce
the lines that prompt the crossing-over places in colour.
This results in hexagons, star hexagons and 60°, 30° rhombs.

SIXNESS

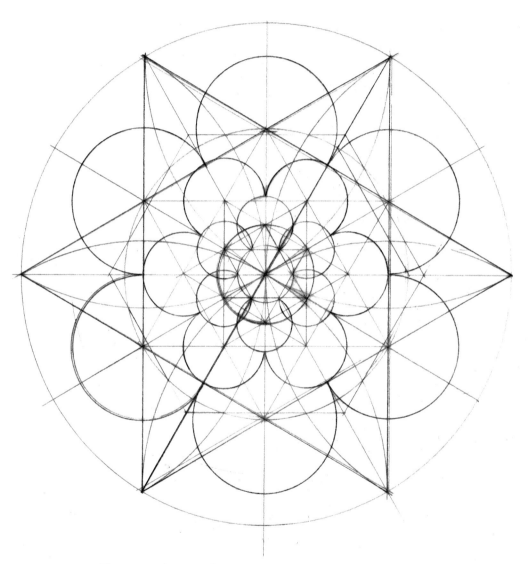

This is an exploration of sixness in circles diminishing proportionally.

The centres of the geometric petals are given in this proportional exploration.
The first is our initial trial, the second with firm lines.

SEVENNESS

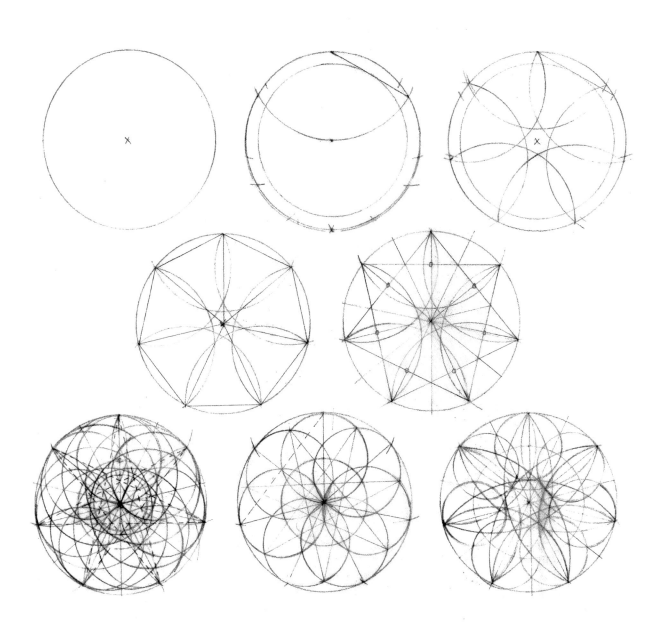

This page goes through stages of the relative construction of the heptagon or sevenness. The first two drawings demonstrate a good way to achieve a very close sevenness inside a given circle. The remainder of the drawings are elaborations and flowerings.
We have kept our construction lines for the serious student or researcher.

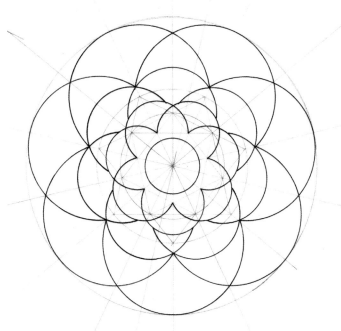

An exploration of sevenness in proportional circles and arcs.
This can be taken as a geometric seven flower.

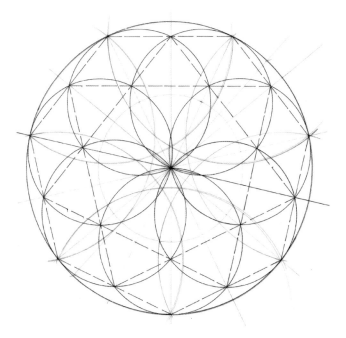

Another aspect of the geometrical flowering of sevenness showing a heptagon, a
star-heptagon and a secondary sevenfold flower, based on seven circles
of half-radius within the parent circle.

SEVENNESS

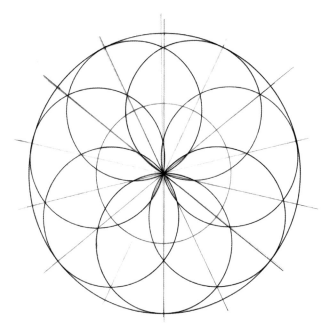

Now we turn to seven half-circles on their own within a parent circle
and then three types of 'petal' become apparent.

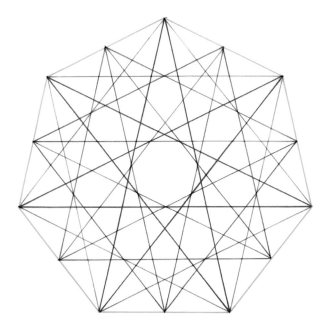

This pattern shows a set of co-ordinated seven-pointed stars arranged
concentrically.

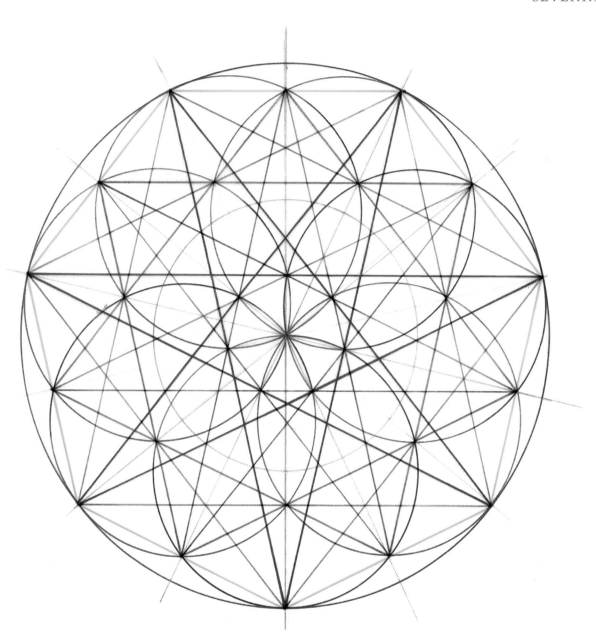

Now we turn to seven half-radius circles within a parent circle. We also find a
beautiful geometric flower in a rare symmetry with three 'petal' types.

EIGHTNESS

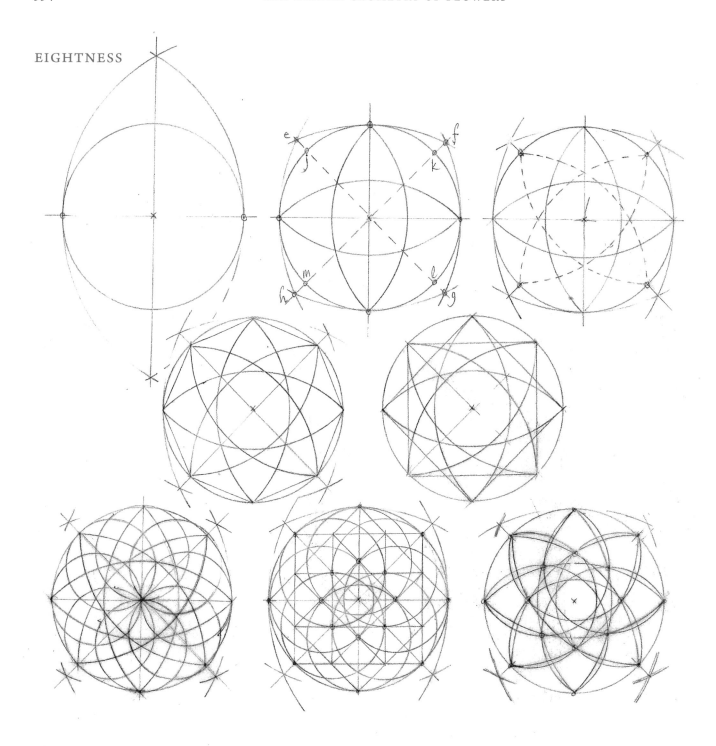

This page explores aspects of the creation in geometry of eightness from the single circle onwards by division. Construction lines have been kept for the practitioner.

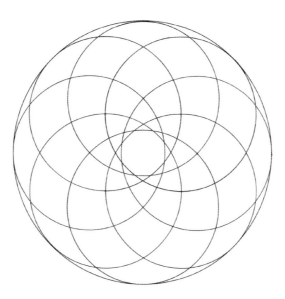

Another exploration of eightness with off-centre overlapping circles demonstrating
the relative freedom within the discipline of geometry.

This leads us to a further exploration in eightfold arcs.

EIGHTNESS

Now we move on to eightness in geometrical flowering.
These are the eight half-radius smaller circles symmetrically arranged in the
larger parent circle. There is a beautiful precision as to how all eight meet at
the centre. Three 'petals' arise quite naturally from this diagram.

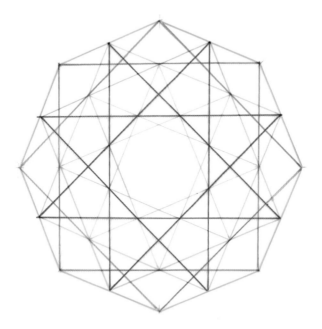

A set of concentrically-arranged eight-pointed stars.

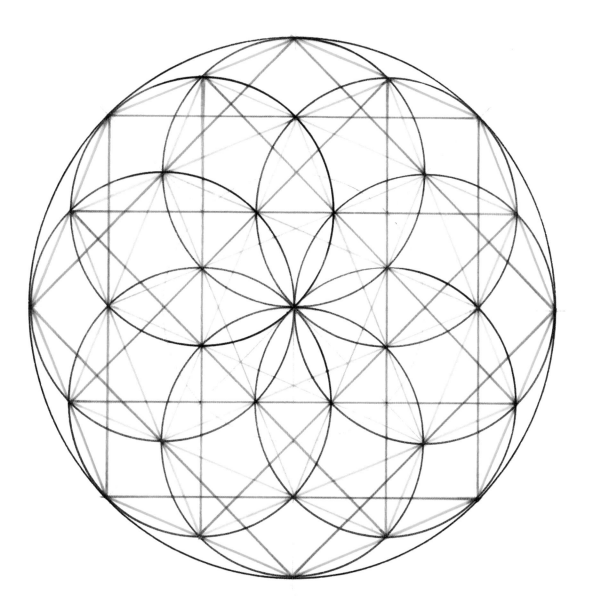

Here we show the lines of crossing points in colour, and
both drawings in concert.

EIGHTNESS

A further exploration into different eightfold
petals proportionally related . . .

. . . and paired with a speculative mix
of straight lines, circles and petals.

EIGHTNESS

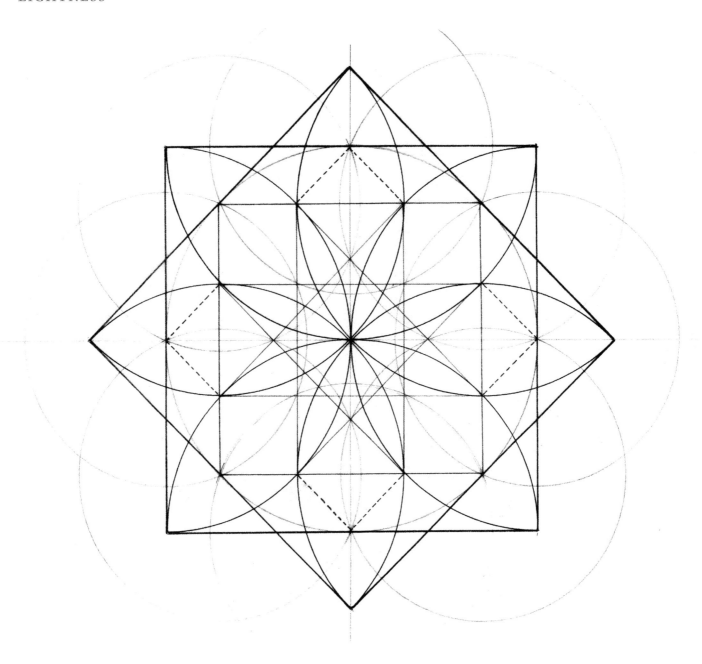

Another aspect of an eight-petalled 'flower' and the related 'square' proportions.

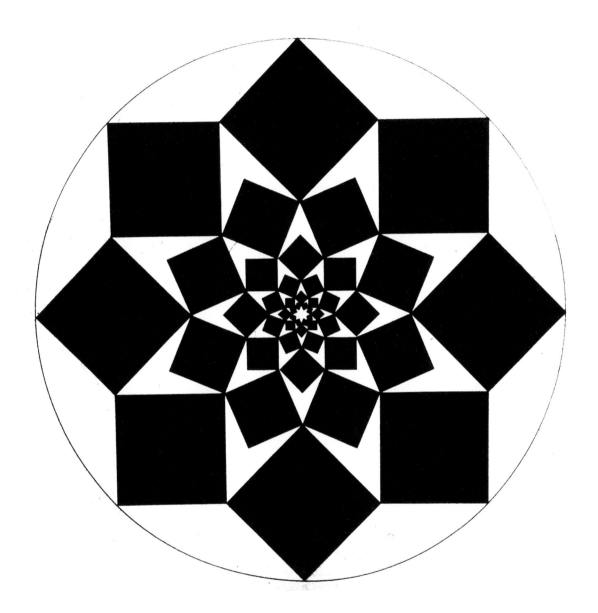

In our final consideration of eightness we show the 'rings' of diminishing squares.
Each symmetry can be represented in this manner, but we choose
only eightness here. Note the 'white' star at each level.

NINENESS

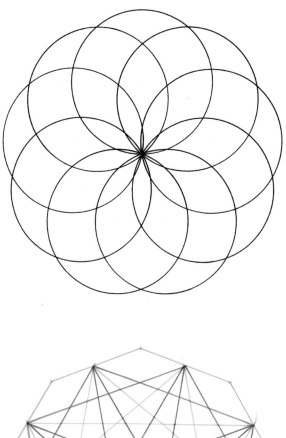

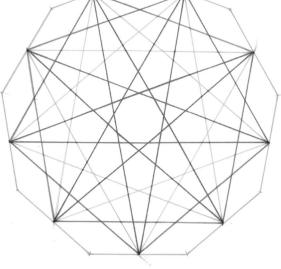

To be consistent we now turn to nineness. Here is the archetypal pattern of nine
circles crossing. In this instance we have not enclosed them in a larger parent circle.

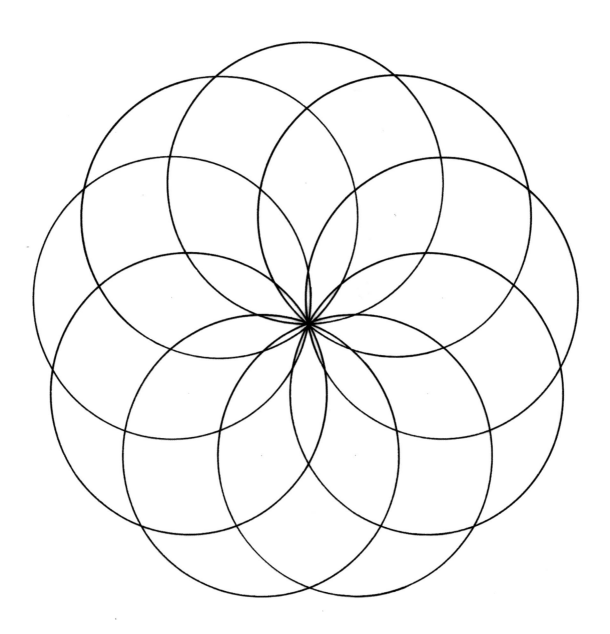

Straight line version which aligns with the circles' crossing-over places.

NINENESS

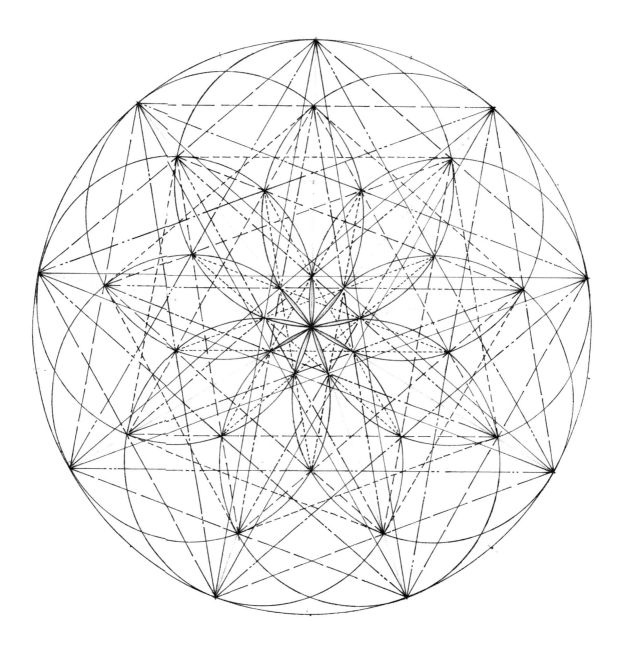

The nine set in both straight lines and circles.

A pair of nine-petalled geometric 'flowers'.

NINENESS

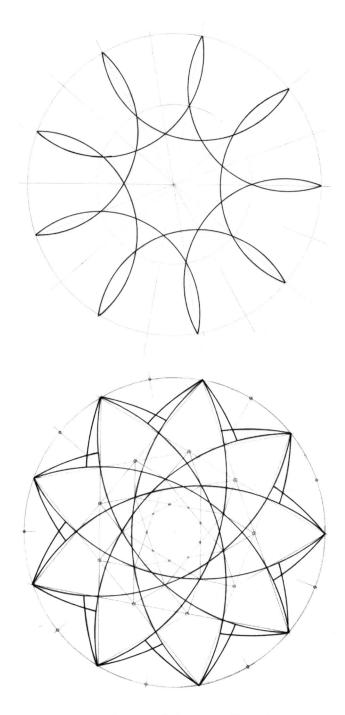

A pair of nine-petalled geometric 'flowers'.

Another pair of nine-petalled geometric 'flowers'.

TENNESS

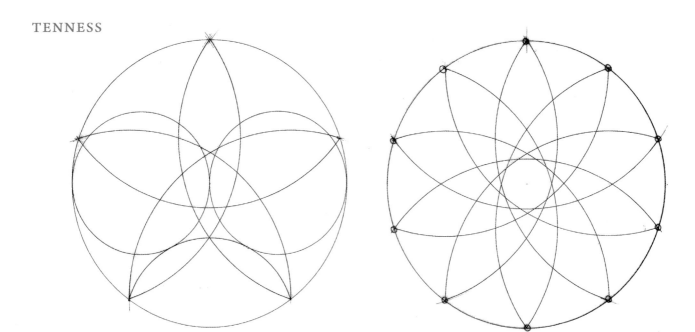

The pentacle.   The double pentacle.

Two ten-petalled 'flowers' in co-ordination.

Ten circles all touching the centre point systematically,
five in red and five in black.

Note that the crossing-over places coincide.

TENNESS

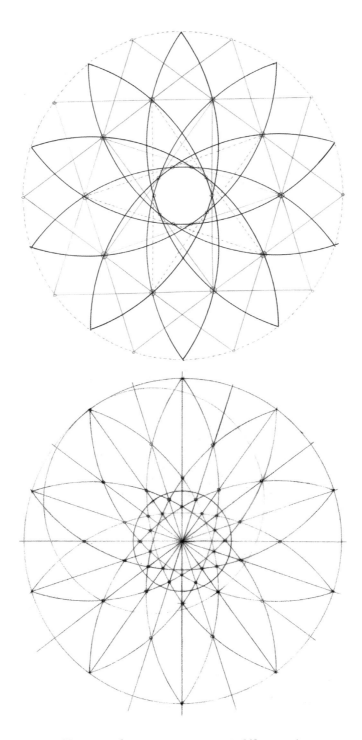

Now we explore tenness in geometrical 'flowering'.
Petal apices emphasized for clarity.

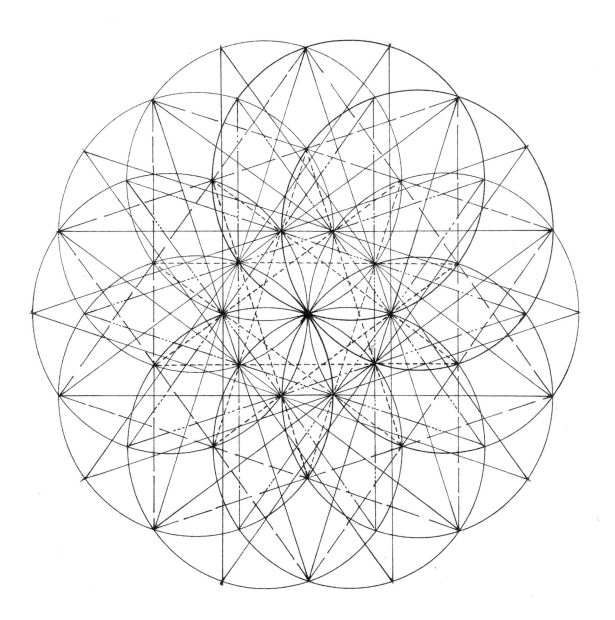

Exploring many more inner relationships of tenness.

THIRTEEN

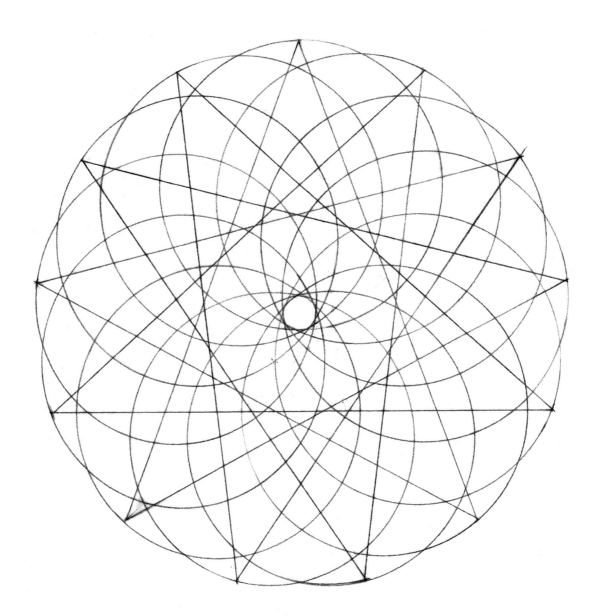

We choose now to move to thirteenness as it is in the golden number sequence.
In this drawing the straight lines do not coincide with the circles' crossings.

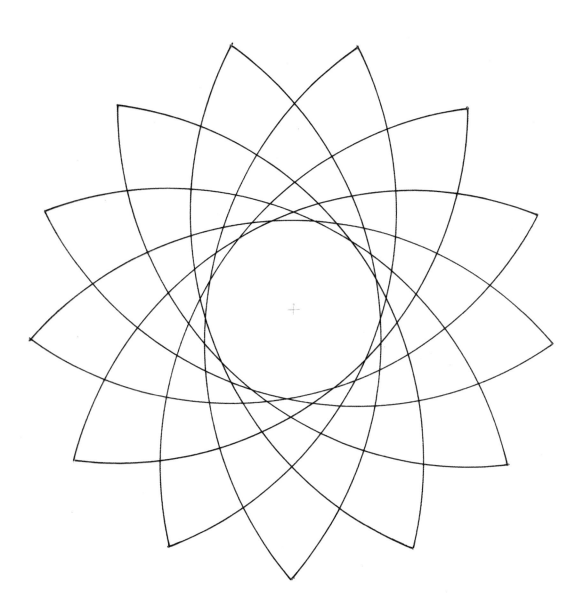

A thirteen-petalled 'flower'.

TWENTY-ONE

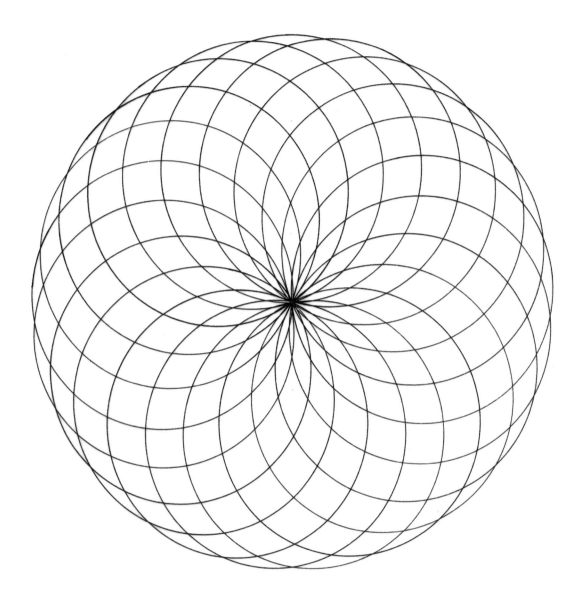

From thirteen we move to the next number in the golden series: twenty-one.

Here are four examples of twenty-one-petalled flowers.

The Internal Divisions of Unity

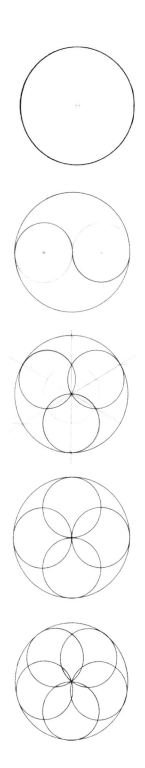

ONENESS:

1. Visible periphery.
2. Invisible centre point.
3. Interspace including all radii.
One is One and All alone and ever more shall be so.

TWONESS:

Merely touching, no overlap Yin-Yang Tao.

THREENESS:

Three overlaps.
Tenness overall in separate shapes.
Creates the completion of the beginning, middle and end of all things.
The spirit moves in a threefold way initiating, sustaining and returning.

FOURNESS:

A double twoness at right angles.
Thirteen shapes. Four is the number of worldly stability: the elements of matter, the directions, the seasons and stages in the life of humans.

FIVENESS:

The wonder of the pentagram and its golden ratios.
The signal of the living.
Five is a handful. The emergence of the golden ratio and the harmonies of the pentatonic musical scale.
Five extensions to our bodies, five toes and five fingers to each limb.
Five alive.

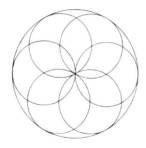

SIXNESS:

The perfection of the crystalline state.
Six is the perfect number of 1+2+3=6 and 1x2x3=6
Days of the creation.
The flowering of the mineral world.
Crystallization.

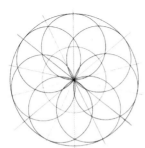

SEVENNESS:

The contained mystery of Seven.
The virgin number and virgin polygon.
Seven Heavens. The number of the visible planets which
are the 'Gods' traditionally.
The notes within the octave.
The number of Athena, Mary and the mystery of being the
virgin number.

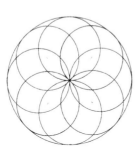

EIGHTNESS:

The completion of the diatonic scale.
The cosmic bodies + mother earth.
Completes the octave and the octagon.
Dynamic and passive fourness together.
The four experiences of heat, cold, moisture and dryness.
Elements of fire, air, water and earth.

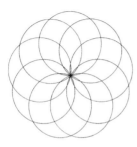

NINENESS:

Nine completes the heavens: visible planets + sun &
moon.
The triple of the spirit.
The highest number before the return to the unity of ten
and the subsequent repetition of the cardinal numbers.

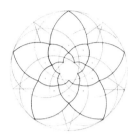

FIVE

LEAVES AND LIFE'S
MOST CONSISTENT MIRACLE:
PHOTOSYNTHESIS

Upon the face of every green leaf
For the people of perspicacity
Is the wisdom of the Creator.

Quoted by Dr S. H. Nasr from a Persian Sufi Poem [48]

The disposition of sunlight-feeding leaves arising out of the same stalk or twig
is to do so in a neighbourly manner so as not to steal another's light.
This is called phyllotaxis and has a series of strategies.

The Leaf

From start to finish, the plant is nothing but a leaf.

Goethe

Suppose that a scientist wants to understand nature. He may begin by studying a leaf.

Masanobu Fukuoka

As with the flowers themselves, we will look at the leaf in a general way as well as in a particular way. Naturally there are very many kinds of leaves with a huge variety of forms. These all have certain physiological functions in common. Leaf distribution or phyllotaxis helps us understand how the plant, bush or tree chooses to let light through. The 'ins' and 'outs' of the leaf forms relate to each leaf's 'love of its neighbours'. Life is always collaborative as well as sacrificial, and not merely competitive or confrontational.

Once we have acknowledged the physiological function of 'gaining' light and letting it pass between, we are free to look at their geometry. If the *only* function of a leaf was to act photosynthetically (transforming light into nourishment) they would all be the same shape, which they patently are not! So we can allow ourselves, out of respect for the leaf-producing plant world, to look at another dimension of this miraculous world of light transformation. After all, photosynthesis is the mastery of sustaining life and enlightenment at all levels, symbolically and actually.

We will take only a few leaf types and show some of the remarkable geometrical coincidences in their structure. Explanations are far from obvious or simple. We are only making observations which respect the seniority of the plant world and maybe reveal some of their intrinsic geometry.

When we have explored this aspect of the leaves we will also look at the other end of the plant world's purpose — which is to produce fruit and seeds. Fruits and their seeds quite obviously have a profound singularity about them which we will also touch upon. Because we take fruit for granted as food we tend to lose our wonderment at the fact that a plum tree so consistently produces plums, and apple trees so reliably produce apples, and so on. These are differences that matter profoundly.

A flowering plant is inevitably a whole: roots, stalk, leaves, flowers and

Transluscent leaf.

Top. The Passion Flower (*Passiflora*) has a beautiful orange fruit.

The Rosehip or seedpod is also a source of intense vitamin C for human beings and others. The author knew its benefits as a child during the last world war.

fruit. As a one-time student of art, the author was encouraged not only to 'see' accurately with a sharply focused vision, but also taught how to screw up one's eyes so that one is able to see the *whole* of any vision in front of the eyes. This technique has lasted, and is a beautiful frame for the discipline of seeing the whole as well as being clear about the details. We have a huge field of vision but we only focus on a single part of it at a time. This can be related to both 'field' and 'point detail', as they are different perceptible perspectives both factual and metaphorical, actual and symbolic.

It is surprising how few have taken the time to discipline their eyes to find out what sight really *can* be, whereas many train their ears to be musical.

To *see* a tree as a whole is a most valuable experience, although one has to stand well back from most trees so as to set their whole form within one's sight. How many of us have been struck by noticing how the overall profile of a given tree is remarkably similar to the overall profile of one of its leaves? This is not true of all trees but of some in particular.

There is obviously a huge variety of leaf types, yet there is no theory known to the author that can explain this beautiful and magnificent multiplicity of shapes and sizes. All leaves are part of the plant's circulatory system and the vehicle of photosynthesis. This is the miracle that is so easily overlooked. Without photosynthesis there is no life on earth. As the Judaeo-Christian Bible states, 'All flesh is grass'.

> What shall I cry? All flesh is grass, and all the goodliness thereof is as the flower of the field.
> The grass withereth, the flower fadeth . . . but the word of our God shall stand for ever. (Isaiah 40:6-8)

(Reportedly Albert Einstein's favourite Biblical quotation).

We can equally say *all* grass, or green stuff, is the transformation of light. Is this not equally a miracle? It is thanks to leaves and seeds that we are alive and have food.

The word 'leaf' is the ultimate term of reference for flatness. We call a thin sheet of any material a 'leaf': a leaf of paper, a leaf of gold or the same of silver — even minerals, or stone are cut into leaves at times. Most metals can be reduced to 'leaf' thinness. Pages of a book — even as you read now — are carrying their messages on leaves of paper. Books were originally written on leaves and in some traditions still are. All true papers are plant fibres.

Although leaves are singular in their main feature of 'curved' flatness, they come in a remarkable array of patterns, from the most simple or archetypal called *ovate* or even *acuminate*. This can be symbolized by the essentially simple form of two intersecting arcs called the *vesica*. This form is so primary that it figures most notably as Euclid's first proposition of how to create an equilateral triangle. Thus with the *vesica* the primacy of the leaf is closely related to the primacy of the √3 (root three) proportion. (The root three proportion is that of the width of a *vesica* to its height. The square root of three is a number that evades a final whole number solution. Plato's Timaeus suggested that it was the most beautiful triangle in the created universe — the half equilateral triangle.)

This brings us to the naming of leaves. Naturally this has its problems, both of convention and appropriateness. So, we are given lanceolate, sinuate, palmately lobed, hastate, reinform, crenate, cordate, serrate, obtuse, perfoliate, connate, trifoliate, palmate, alternate, pinnate, bipinnate, whorled, lingual and so on. At this point the author challenges the non-specialist to visualize the shapes of the leaves that each name is describing. Even the specialist may not find it easy to bring all of these leaf types to mind. What we can say is that the image of a leaf such as that of an oak, if well burned into the memory, will be of greater value to the memory than learning the conventional name (although obviously both

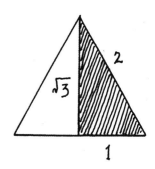

This triangle has the 'ineffable' quality of the square root of three as one of its sides. As we have repeated, this triangle holds immense importance in Plato's cosmogony.

An analysis of a tree form using an equilateral triangle and the *vesica* motif. Leaves from certain trees have profiles quite similar to the whole tree.

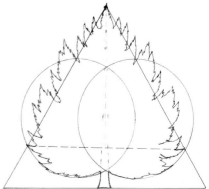

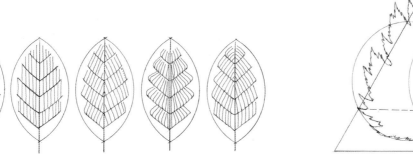

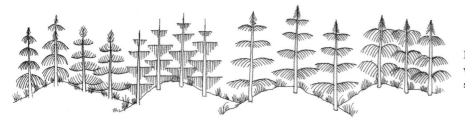

From the venerable Tibetan art tradition we find a most interesting array of symbolically-represented leaves and trees.

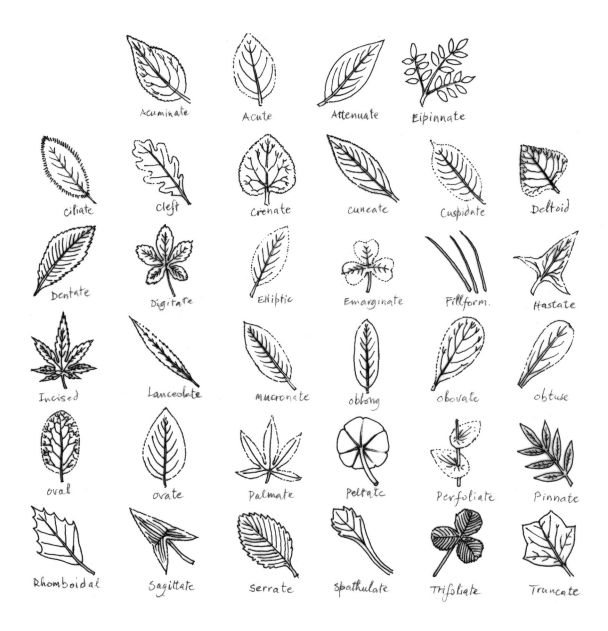

The great variety of leaf types demonstrates that photosynthesis is not the only function of a leaf. As yet there is no consensus about the names of leaf types, but here is one naming convention. But the question remains: why is there such immense variety?

are valuable). As the great Muslim poet Rumi asked, 'Whoever plucked a rose from R - O - S - E?' This must not, however, diminish the importance of names or confuse the difference between 'reality' and 'actuality'. Recognition is the ethos, all else is to this end. Or to quote another master of poetry, Kabir said:

> Because someone has made up the word wave,
> Do I need to distinguish it from water?

While considering the miraculous beauty and variety of leaves, our present task is to draw attention to the equally profound aspect of their inner as well as their profile geometry. There are certain key angles that recur in the study of leaves that have no simple explanation. These can be related to the outer shape of the leaves, or to the veins within them. Both yield unexplainable results when carefully studied. In a manner similar to flowers, leaves show an aspect that can be bilateral, trilateral or more in symmetry. Singular leaves appear in the grasses and are called monocotyledonous (a word that is anything but simple). However we also find an example of singularity in the *perfoliate* leaf, where the stalk passes right through the singular leaf form. Most leaves are bilateral on either side of their midrib, but the beautiful exceptions to this pattern are the *sinuate*, the *palmately lobed*, the *palmate* and the *pinnate* families, which make up secondary patterns containing their individual separate smaller leaves. We will look into this more deeply in illustrative form, as this 'outline' aspect may hold evidence of an invisible or intangible aspect of leaf formation.

Leaves close to the classical *vesica* form. Near and far as sameness and otherness.

An aspect of leaves that Goethe drew attention to (which has not received sufficient attention in the author's view) is that of the changes in leaf shape at different levels of the plant above the ground. This has been well treated by followers of Goethean science and researchers of the Anthroposophical Society.[49]

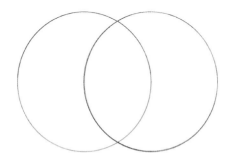

The *vesica* itself.

This is known as a Kirlian image of a leaf. It somehow seems to confirm the findings of
Sir Jagadish Chunder Bose, and certainly hints at the subtle envelopes of the living which
are fundamental to the Vedic perspective.

The Leaf: A Definition of Its Nature

There is no life-reaction in even the highest animal which has not been foreshadowed in the life of the plant.

Sir Jagadish Chunder Bose

In the courtyard there grows a strange tree
Its green leaves ooze with a fragrant moisture.
Holding the branch I cut a flower from the tree
Meaning to send it away to the person I love.

Anonymous, Chinese lady poet (translated by *Arthur Waley*)

Geometrically, a leaf is a flat plane — a thin green lamina or blade made mostly from a soft tissue of thin and translucent walled cells. The central line of each part of a leaf is called its midrib. The lines branching from the midrib are called the veins or nervure. The majority of the leaf's surface is called its blade.

The whole leaf surface is supported by the thicker and stronger pattern of veins. The stalk of the leaf, also known as the petiole, continues into the midrib or main vein. Sometimes however there is no leaf stalk, so the leaf emerges directly from the 'twig' or even a larger part of the plant. Naturally, the leaf thickens, particularly along the midrib, to give structural rigidity, but also to carry the nutrients and liquids both ways.

Leaves normally 'read' beautifully as wholes. For anyone taking time to contemplate it, a leaf will always be rewarding — unless you take hold of a Stinging Nettle by mistake.

It is particularly interesting to observe how leaves arrange themselves along the stems of their parent plant. This is by no means a standard or mechanical function, although it has an obvious yet subtle value of ensuring that the leaves work collaboratively and each takes as little light from its immediate neighbours as possible. The emergence of the leaves in sequence along the stems is called phyllotaxis. The geometry of phyllotaxis repeats the archetypal possibilities of universal space distribution, with the natural references to particular angles. (See the Nasturtium leaf study on page 369.)

It would be sad in our intensity of study to miss the immense beauty

How the Nasturtium Speaks to Us.

An 'angle' is a 'gon', a 'generator' in the original Greek. So can we assume that, according to the viewpoint of Proclus and the sages of the Greek tradition, each angle (or portion of 360°) is an 'intelligence' in itself and thereby a generator? In the green world of plant life we find particular angles recurring, the most persistent of these being 90°, 45° and 30° on the spatulate form of the leaf. Leaves offer themselves to the light in the same way as flowers and consequently offer themselves to the attention of humans and other animals. Thus if we entertain the idea that there is more to the plant world than just food for animals and occasional sentimental delight to humans, then we have to pay attention to the language of geometry that it is offering us. It stares us in the face if we are awake to it.

that not only lies in the differences between exact neighbours on the same stem, but also in the differences of greenness, form, veining, colour changing and wind-caused movement of leaves. Every good gardener knows how to complement the different greens when planning a beautiful garden layout. Vegetable growers know how important greenness is and the huge variety of greens within the simple term 'green'. The autumn colours of leaves are often breathtakingly beautiful.

In order to work, photosynthesis requires the assistance of a supply of carbon dioxide or what we, as animals, breathe out. Thus the leaf is also a breathing vehicle as well as a liquid one. The flatness and breadth of the leaf facilitate rapid absorption of both oxygen and carbon dioxide while also encouraging the maximum amount of sunlight to fall on its surface. This breadth ensures that the necessary gases are transported the shortest possible distance. While the gases are being exchanged the same facility encourages the evaporation of water as vapour. This is called transpiration and is a vital aspect of collective leaf behaviour — and one of the most mysterious functions of plant and tree life because it is taking water *up* on a journey quite the opposite to water's natural tendency. The transpiration of trees even causes cloud formation above a forest. We all

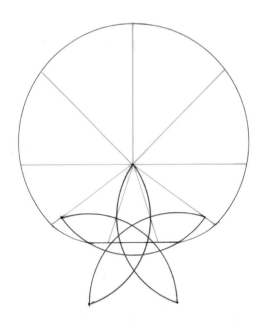

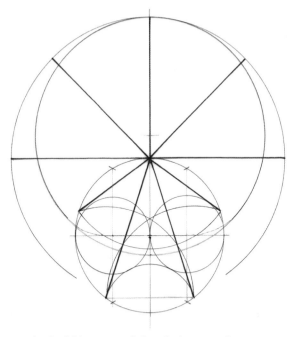

1. We have taken the nine veins radiating from the centre of the Nasturtium leaf as the basis for suggesting a geometry of four equal angles above the horizon line and five below.

2. The fivefold geometry below the horizon takes its centre from the base of the leaf.

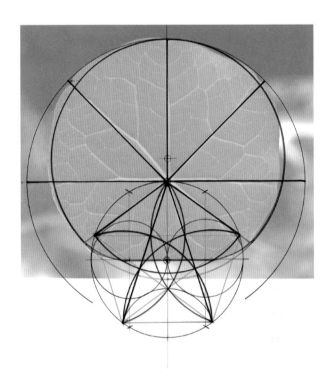

3. This fivefold symmetry adequately follows the four veins radiating below the horizon.

This essential molecule of light synthesis has a wonderfully mandala-like symmetry. It comes in two subtly different patterns.

know that water 'pools' as low as it can under normal circumstances. Life, it seems, reverses water's pooling character.

The thin leaf has its special arrangement of epidermis or 'skin' cells which fit closely with no air spaces between. The outer skin covers itself with a waxy substance, not unlike greaseproof paper, and this 'cuticle' layer controls evaporation as well as performing many protective functions and helping to keep the leaf in shape and translucent. The sunlight passes directly through these layers to the next layer of cells which contains the *chloroplasts* which perform the essential function of transforming light into food. Exactly *how* they achieve this is still a mystery — a mystery enhanced by the magnificent chlorophyll molecule *with its mandala-like form.*

There is delight in discovering that chloroplasts are the light-absorbing element in cells which are typically lens-shaped and bounded by a double membrane. Part of the structure of chloroplasts is found in their thylakoids. It is in this membrane that the mysterious transformation called photosynthesis actually takes place. (There is a pleasing resonance between *chloros,* the Greek for 'green' in 'chlorophyll' and 'chloroplast', and 'Chloe' the Greek goddess of flowers.) Respecting this mystery, a good scientific text will say: 'This chemical activity *is believed* to take place on the surface of the chloroplast when it is receiving light'.[50] Chloroplasts are understood to build up sugars and starches prior to their distribution as food to the plant — thus also providing food for the ruminants and creatures who eat green leaves.

There is one more aspect to leaf structure which holds its own mysteries as well as being known to perform certain functions. These are the *stomata* (the singular is *stoma*), described as 'tissue that forms the framework of an organ' (like the tissue of the ovary that surrounds the reproductive cells). *Stomata* exist as a feature of all leaves that are involved in transpiration and breathing or 'gas exchange', as well as water-shedding. They are part of the upper and lower surfaces of virtually all leaves. Although more abundant on the lower side of a leaf, probably due to their water-shedding function, these curious structures are abundant all over the leaves. *Stomata* are openings or, more strictly, structures that can open in the epidermis or 'skin' of any broad leaf. Each stoma is signalled and set up between two guard cells. These 'side' cells were believed to work according to their internal water pressure or *turgor,* but recent studies have shown that this is not the case — we still do not know exactly what opens and closes the stomata. The primary cause is thought to be principally connected with light intensity but we cannot be sure. The mechanism is understood to be a chain of events that leads to the increase in the concentration of sugars in the cell's sap in the interior hollows (or vacuoles) of the side guard cells.

Eventually the walls of the guard cells swell and this in turn causes the outer walls to stretch, making the guard cells curve away from each other. This facilitates an opening between them, not dissimilar to a mouth or eye opening.

This curious organ-like structure poses an mystery as it is associated with gases, liquids and (to a degree) light intensity, and has the quality of proto-nostrils, proto-lips, and proto-eyes.

When faced with the intense subtlety of what is continually going on in the body of a single leaf, one senses the limitations to the word 'intelligence' or at least our current human definition of intelligence.

Again we refer back to the important work of Sir J. C. Bose of Calcutta in the 1920s. He showed that plants have responses that are similar to those of an animal nervous system. The open-minded scientist will agree that 'intelligence' and 'consciousness' are still very much on the agenda when it comes to definitions and understandings.[51]

This inspired diagram shows the remarkable changes in the shape of trees at different elevations. (With acknowledgments to Viktor Schauberger)

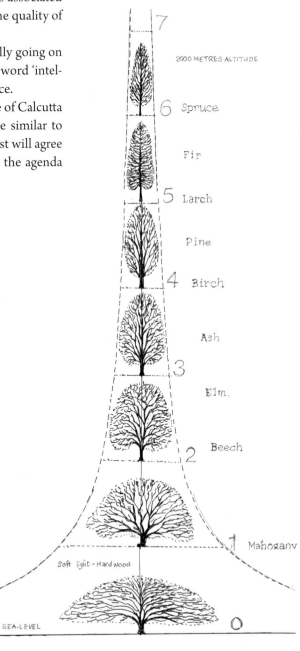

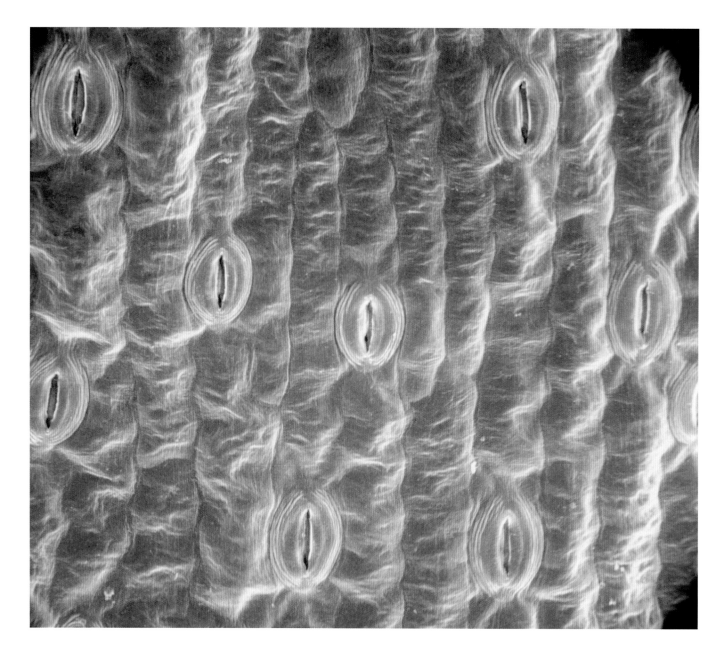

Stoma distribution on the underside of a typical leaf.

Photosynthesis: What Exactly Is It?

I turned to the light — and never looked back.

Anon

And of light, itself incorporeal, the culmination or flower so to speak is the sun's rays.

Emperor Julian

Photosynthesis is surely as fine a metaphor for 'the bringing down of light' as any within our human comprehension. That we take this miracle for granted is a reflection of the attitude that pervades the modern world, and that encourages us to take life itself for granted, and in doing so we lose all reverence and respect for it. This constant flow of light into lower workable forms of energy is essential to life. If we look at the perennial philosophical premise that consciousness flows through four distinct worlds, levels or degrees, from subtlety to grossness, we find that light is, both metaphorically and factually, the most subtle and the finest, and that it is essentially the source of all 'thingness' in the material domain when defined as energy. The four distinct worlds are the 'states' of matter traditionally called fire, air, water and earth and each represents a threshold: the radiant state, the gaseous state, the liquid state and the solid state. These four 'states' are the material expressions that Titus Burckhardt,[52] as spokesman for the perennial philosophy, speaks of as metaphors for the human levels of consciousness.

Thus, if we may be permitted, *light* is inspiration itself. *Air* is the established domain of the Intellect, which becomes the foundation of culture. *Water* symbolizes the emotive relationship between people and all living things — the social level. Finally, the *solid* state symbolizes our experiences of materiality or the sensorial world of 'food', tangibility and physical things, the necessary support for all physical bodies.

Thus photosynthesis performs the primary vital task, of bringing the higher level of these four stages right down into the lowest levels of working fuel or physical food. This can be expressed chemically, yet in so doing we avoid recognition that it then appears only materially or mechanically, and we miss the extraordinary feat that the cells in any leaf's surface perform during all daylight hours. A leaf is *living* chemistry.

Not least of the functions of the 'green' world is the production of

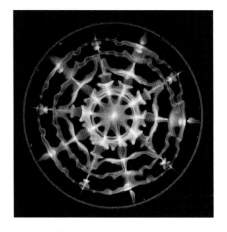

The tenfold vibration pattern arising out of the experimental work of Lauterwasser in Germany demonstrates how sound creates form in the sensitive medium of water.

oxygen as a necessary by-product of the action of photosynthesis. The reciprocity of carbon dioxide (CO_2) and oxygen (O_2) is again a beautiful metaphor for the symbiosis, or complementary breathing relationship, between the vegetal kingdom and the animal kingdom.

Traditionally, bodily life was spoken of as having to 'bear' or 'harness' the four elements from which all things are fabricated. For the sages of ancient times in all traditions, the synthesis between mind and matter was the responsibility of philosophy. In this sense as much effort was focused on what is 'out there' as was focused on the self, and on why our self believes there is an 'out there'. Life harnesses the four elements and harmonizes them for its own purposes. We take in the outer world and in doing so retain its impressions and make them into ourselves. It has to be seen as a profound interdependence.

Water quite clearly is the vehicle of life with a remarkable flexibility which can accommodate the greatest number of different forms. The result of introducing vibration into still water at different frequencies demonstrates this admirably.[53] The cause of the resultant form indicates its source in a lighter or more subtle level. Traditionally this subtle level leads the logical mind from the world of things to the domain of principles. It is clear that the finer is influencing the more dense yet both are inextricably linked. Proportion and frequency are thus seen as equally integral. Socrates' line can be seen as the symbol of continual connectedness.

The miracle of sunlight synthesis. The tiny little 'valves' called stomata, usually on the underside of a leaf, are proto eyes, proto nostrils and proto mouths, even though they are minute. They have many different functions for the leaf and plant.

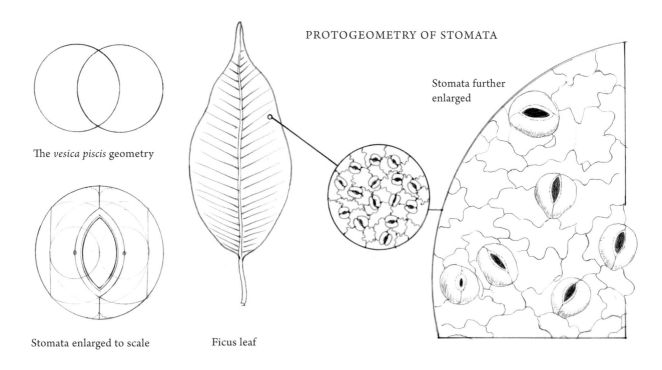

PROTOGEOMETRY OF STOMATA

The *vesica piscis* geometry

Stomata further enlarged

Stomata enlarged to scale

Ficus leaf

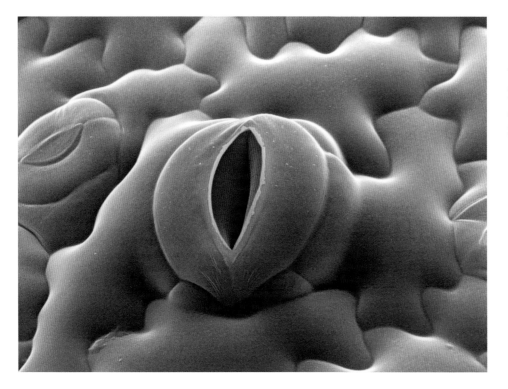

This close-up image of a single stoma demonstrates its striking beauty and vivid colouring. The stomata breathe, weep and let in light.

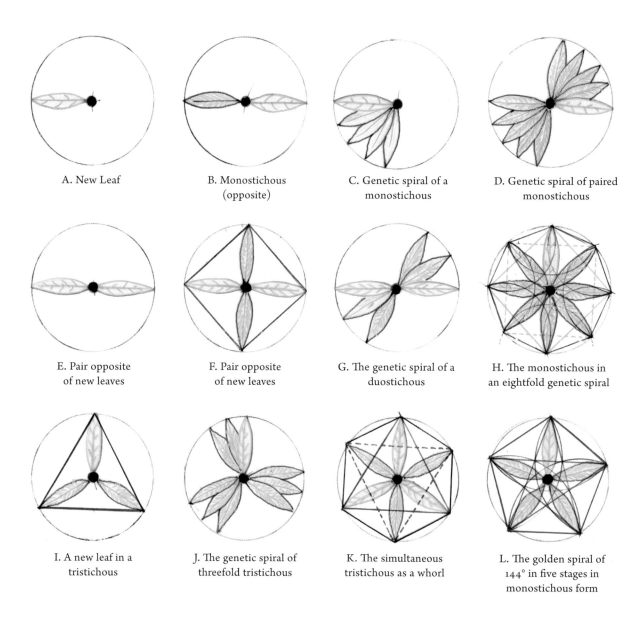

A. New Leaf

B. Monostichous
(opposite)

C. Genetic spiral of a
monostichous

D. Genetic spiral of paired
monostichous

E. Pair opposite
of new leaves

F. Pair opposite
of new leaves

G. The genetic spiral of a
duostichous

H. The monostichous in
an eightfold genetic spiral

I. A new leaf in a
tristichous

J. The genetic spiral of
threefold tristichous

K. The simultaneous
tristichous as a whorl

L. The golden spiral of
144° in five stages in
monostichous form

PHYLLOTAXIS

Leaves distribute themselves so as to help the other leaves receive their share of light.
This is in the interest of the whole organism.

Phyllotaxis or the Display of Leaves on their Parent Stem

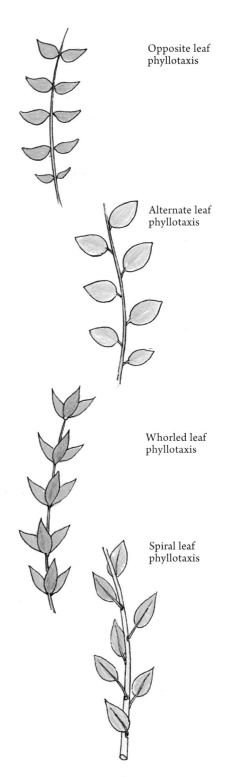

Opposite leaf phyllotaxis

Alternate leaf phyllotaxis

Whorled leaf phyllotaxis

Spiral leaf phyllotaxis

Leaves are silences around which flowers are their words.

Rabindranath Tagore

Each laughing rose that springs from his lips
Has avoided the thorn of existences,
Has escaped the sword of corruption.
Every tree and plant in the meadow seems to be dancing,
Those which average eyes would see as fixed and still.

Rumi[54]

Phyllotaxis is the choreography of leaf patterning.

K. B. C.

The life and sustenance of any plant, whether Herb Robert or an Oak tree, depends on the display and light-absorbency of its leaves. The word 'phyllotaxis' derives from the ancient Greek *phyllo* meaning 'leaf' and *taxis* meaning 'orientation in movement'. Leaves display a remarkable number of choices as to how they orient themselves. Naturally the primary motivation is to receive as much light as possible yet to avoid interfering with the light-needs of the neighbouring leaves. Leonardo da Vinci also observed that they were disposed to maximize the gathering of dew, which also points to their rain-shedding role.

Needless to say, it has taken literally billions of years, *well before we as humans arrived on earth,* for the plant world to explore and express all of the phyllotaxic strategies.

The first geometry that needs to be investigated is that of the stems of branches that are going to be leaf-bearing (with the stomata located mostly on the underside of the leaf).

A single 'specimen' tree in an open site will naturally tend toward a spherical or semi-spherical form. This is despite the fact that direct sunlight will only be available on three sides during any particular day. Some leaves on the north side of a tree will have to rely on light from the sky rather than from direct sunlight. However we must not forget that each leaf is able to 'dance' in the breeze on its stalk, thereby gaining

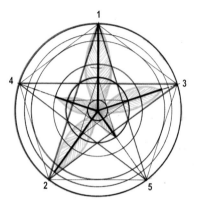

A. The full range of proportions that
determine this leaf growth pattern.

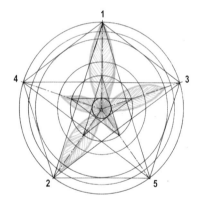

B. This drawing demonstrates qualities of
the pentagonal symmetry determined by the
oldest leaf (topmost 1).

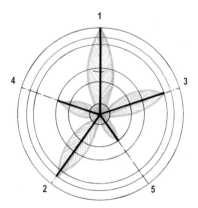

C. The ages and emergence of the leaves
according to the golden series in fivefold
symmetry.

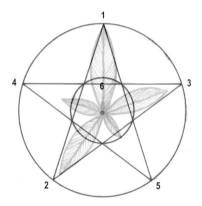

D. The two circles here represent
the greater or oldest leaf as radius
and the smaller or fifth leaf as radius.
This is a special proportion.

fuller orientations towards the light. The rustle of leaves, and the tones that the wind makes as it passes through trees, are often a valuable part of childhood memory for those of us who have been fortunate enough to experience them. Trees also hum, chant and sing to the imaginative and sensitive.

The choice of rhythm by the leaves sprouting from the stem of their plant has an overall effect on the total shape of the shrub or even the tree and its branching. Branching is itself another beautiful geometric study.

The point at which a leaf grows out from the stem is called a 'node'. How many leaves arise at each node is the first categorization of phyllotaxis. The three principal categories are 'alternate', 'opposite' and 'whorled'. The fourth category, 'spiralled', can apply to each of the first three.

One leaf per node is called monostichous (see illustration A on p. 376), two leaves per node are called distichous and three leaves per node are tristichous. The effect of spiral distribution in time, as the stem grows and produces the next node, is governed by geometrical rules. This spiral effect is called spiromonostichous in groups such as the three-leafed.

The spiralling effect helps the plant as a whole. As each set of leaves increases up the stem, particular angles are employed which tend to reflect a 'golden' progression numerically and the angles tend towards the ultimate golden 137° 30' 28" *without actually getting there*!

The spiralling effect once again reveals the golden series or the Fibonacci numbers such as 1, 1, 2, 3, 5, 8, 13, 21 etc. as an overall system, yet each plant type has its own chosen angle or angles.

Alternate leaves will have an angle of half a full rotation, this being 180° as it is half of the full 360°. In the Beech and Hazel trees we find the angle of winding per leaf to be one third of a full rotation — thus 120°. In the Oak and the Apricot it is two fifths of a full rotation and thereby 144°. (This being two times the 72° of a pentagonal star triangle of 72°, 72°, 36° which embodies the golden relationship.)

The Poplar and the Pear winding angle is three-eighths or 135°. After these come the Willow and Almond trees which spiral a proportion of five thirteenths of a full rotation, producing an angle of 138° 28' . Further fractions of the full circle can be found in other phyllotaxic systems in nature, but we have enough here to demonstrate how the golden series is clearly evident in the plant world when leaves choose to leave the stem at a proportionate rate.

It has been claimed that these patterns of leaf unwindings or spiralling emergence are controlled by the plant hormone auxin.[55] One can only speculate that the good hormone *auxin* was educated at a Fibonacci

In this instance three leaves come out three at time yet they also turn enough so as not to be exactly above their neighbours. This confirms the collaborative thesis.

Proportional unfolding is facilitated by a spiral path. Two examples from the domain of plants.

'Master Diagram' from *A Little Book of Coincidence* by John Martineau. John Martineau made his remarkable discoveries of the proportions of the mean-orbits of the planets in our solar system while a student in the Visual Islamic and Traditional Arts Department of the Royal College of Arts, London, under the direction of the author. His *A Book of Coincidence* (title of the first edition) may well turn out to be of historic significance.

School, as she has a handful of golden strategies to play with. It seems far more likely that there are a series of interactive collaborative 'causes' behind such a significant display of a universal law. To reduce it to a single hormone seems far too reductionist to the author.

We have decided to demonstrate the value of a two-fifths spiral phyllotaxis with five orthostichies (an orthostichy is an arrangement of leaves one directly above another on the stem). This means that the whole spiral strategy is contained within a fivefold symmetry — in our illustration (on page 378) we have named the upright direction as number one. This will represent the oldest leaf as well as the sixth and youngest leaf direction. This genetic spiral (as it is called) can be seen to have passed twice around the stem axis (azimuth) producing the two-fifths fraction and an indirect spiralling measure of 144° (between any two successive leaves). Illustrations, like the leaves themselves, speak far more eloquently.

The symmetry of the containing 'field' of fiveness offers many coincidences of symmetry which we have chosen to illustrate in addition to the layered leaf-growth pattern. It is noticeable that along the axis called number four in our illustration, the distance exactly coincides with the radius of a circle that passes through the inner crossings of the arms of the overall five-pointed star. This is a pattern that consistently emerges in the natural world[56] from the planetary facts to the proportions of the DNA helix seen longitudinally. The author has to thank Adrian D. Bell for his explanations in his most excellently illustrated book *Plant Form*.[57]

There is no better example of an archetype from the intelligible domain influencing life in the sensible or actual domain than the Fibonacci series (the golden proportion). It is no wonder that Kepler, as already mentioned, called it a 'jewel'.

There is a current fashion in the research labs of modern botany to cite 'auto' behaviours in plant form. Phyllotaxis is no exception. It is as if the Darwinian survival of the fittest was an 'auto' pattern of nature, when in fact it is the survival of the fittest to *collaborate* (or *network*) that is the truer picture. The overall meaning of *collaborate* in its broadest application is probably beyond conventional human reasoning.

This section through the long body of the DNA molecule demonstrates an exceptional relationship between the star decagon (ten-sided) and the 'golden' pentacle.

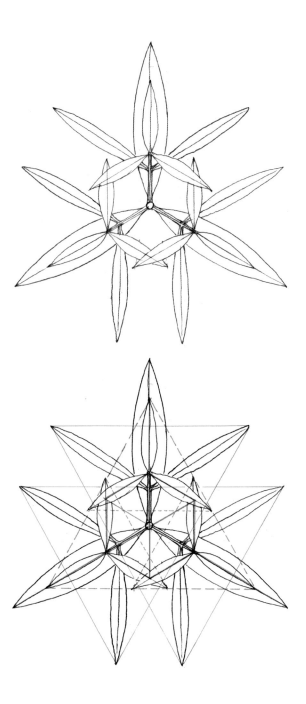

The drawing above is adapted from Adrian Bell's *Plant Form*, page 29.
The intrinsic threeness is most marked in this Oleander twig (*Nerium oleander*)
viewed from above. The archetypal geometry has been suggested
in coloured dashed lines.

Leaves, Chinese Wisdom and Understanding the Recurrent Principles of Plant Life

See how the Orient Dew
Shed from the Bosom of the Morn
Into the blowing roses . . .

Andrew Marvell, from *On a Drop of Dew*

As each light-transforming leaf is the source of the miraculous continuity of life on our planet, the 'green' world should never be underestimated or (one now has to add) abused by mechanical agribusiness and soulless mechanical processing. Financial profit is the least responsible motivation for human behaviour.

Each plant has its own determining pattern from seed to fruits. Even the soil in which it will grow is nourished by the falling leaves, thus demonstrating the pattern to each plant's life cycle and the rhythm of 'green' life. The subtlety and interchange of nutriments is nothing short of amazing in the plant world. It is the epitome of collaboration.

Chu Hsi or Master Chu, the great Chinese sage who synthesized the Confucian and Taoist streams of wisdom, spoke at length on the ultimate mysteriousness of ideas and matter and their inseparable relationship. Ideas being closest to what we have come to call the Platonic archetypes and 'matter' the combined behaviour of the elements or the world of the senses.

We recall that the ancient Chinese honoured 'fiveness' so highly in every aspect of human experience that they settled on the five elements of earth, water, wood, fire and metal (or gold as some like to call it). Master Chu was asked about 'material force' and its relationship to 'principle' (which is the manner in which the ancient Chinese is translated when dealing with 'ideas' and 'matter'): 'Expansion and contraction, and coming and going, are matters of material force. Why did Master Chiang say they are nothing but principle?' Master Chu replied: 'It is according to principle that the material force has to expand or contract and to come or to go'. His reply carefully reintegrates that which had been separated in the question.[58]

A tenth-century wall painting in the Bezeklik Caves in Xinjiang, China, depicts a procession of Uighur people, all carrying above their hearts the flower shown in this detail.

Lotus flowers were also grown in bowls in ancient China. The practice continues today.

Wing-tsit Chan, the eminent contemporary translator and commentator on Master Chu, quotes him as saying, 'Principle (idea) and material force (materiality) must be considered as one'.

Master Chu saw this so-called polarity as an inclusive pulsation: 'In the changes and transformations of Yin and Yang (the passive and active forces), the growth and maturity of things, the interaction of sincerity and insincerity, and the beginning and ending of events, one is the influence and the other the response, succeeding each other in a cycle. This is why they never stop'.[59] Thus there is a paradoxical rhythm within them which Plato called 'Essence' or 'Being' in his *Timaeus* dialogue.

Wing-tsit Chan comments on Master Chu's position by saying: 'It is safe to say that the concept (idea) is there before a principle is demonstrated in specific affairs; it is already present'. This indicates that although principle and material force are integral, principle is prior in the sense that it is always there even before material force is expressed in a form. (Or even if it is not expressed?)

We have chosen to visit the ancient Chinese wisdom tradition as a form of confirmation that the fundamental issues that engage the human mind are universal. Here the closeness to Socratic concepts is quite evident.

This clearly implies that the premise that the geometry of flowers is *prior* to the actuality of flowers themselves is likely to be true.

The Immortals were presented to human gaze by the visionary artists of ancient China.

Symbols and Actualities in the Plant and the World of Cultivation

My friend, all theory is grey and the golden tree of life is green.

Goethe, from *Faust*

If you lose touch with nature you lose touch with humanity.

J. Krishnamurti[60]

No, it is not different,
Now I am old,
The meaning and promise
Of a fragrance that told
Of love to come
To the young and beautiful:
Still it tells
The unageing soul
All that heart desires
For ever is
Its own bliss.

Kathleen Raine, *Lily of the Valley*

A rubbing from an ancient Chinese tomb depicting the sacred art of farming.

In seeking the inner or symbolic meaning within the Great Traditions of humankind we find that for the Judaeo-Christian tradition cultivation is attributed to the third day of creation in Genesis. The goddess of plant cultivation ('culture') was worshipped as Demeter in Ancient Greece and as Ceres by the Romans (and thus 'cereals', the staple foods of humanity).

In most ancient traditions there is an agricultural goddess for all 'husbandmen', farmers and those who 'plough', 'sow', and 'reap' on the earthly plane. But according to Vedic tradition she is *Goddess of the Field of Life*, the *Kshetra* or 'Field' of the Bhagavad-Gita. The *Kshetra* is the entire field of the inner psyche. The 'dweller' in the field is the human soul. In this field the continuous Battle of Life between the Divine Will and the *Titanic-Ahankaric* diabolic forces is always present.

As the *Ahankara* (ego) sows its seeds and thorns, so the 'aspirants for realization' exert themselves in weeding these negatives out. The aspirant proceeds in sowing the good seed and tending and reaping the fruit of immortality.[61] Thus these fundamental activities reflect the different levels of the psyche.

Those engaged in the aspirant's life, knowing that this immortal fruit may be won, are often criticized, even harassed, by those who are ignorant or merely intolerant of the idea of a purpose to life. The Lord Buddha was asked why he didn't cultivate the land rather than beg. He replied:

> O husbandman, I too plough and sow: and having ploughed and sown, I eat. Faith is the seed I sow: good works are the rain that fertilizes it, wisdom and modesty is the plough; my mind is the guiding-rein; I lay hold of the handle of the Dharma [the teachings]; earnestness is the goad I use; and exertion is my draught-ox. This ploughing is ploughing to destroy the weeds of illusion. The harvest it yields is the immortal fruit of Nirvana and thus all sorrow ends.

A solution from the Lord Buddha that does without evil.

The stone carving for the Tent of Peace at St Ethelburga's in the City of London was designed by the author on the principle of simulataneous seed, bud, flower and fruit. It was carved by Adam Williamson at The Prince's School of Traditional Arts, London.

The Cultural Dimension
of Trees and Flowers

That rapture's smile is secret everywhere;
It flows in the wind's breath, in the tree's sap,
Its hued significance blooms in leaves and flowers.

Aurobindo Ghose

Each culture expresses its traditions in its own way, yet the inner meanings reflect our common humanity. First we will visit the Vedic traditional symbolism of India.

On the tenth day of the festival of Durga in honour of the Hindu god Vishnu a buffalo is sacrificed in memory of Durga's triumph over the Buffalo-Demon, after which the delicate *Sami* or *Banni* tree is worshipped. A kind of Mimosa, this tree is sacred to Durga as well as to Saturn (or Sani of Hindu cosmology). This is also the tree of life and is considered very auspicious in the initiation of a child's education on the day of 'Higher Birth' or the 'Birth of Light'. Traditional wisdom is worshipped in the form of doctrines written on the *sacred leaves* or paper formed from the bark or wood of these trees. They are the records of a living tradition, or the necessary tools for the constant renewal of that tradition. (Just as life continually renews itself, so does tradition. This is the re-establishment of eternity when truths drift into passing time.)

In the yoga tradition of the Hindus, there is a scripture called the *Yoga-Vasishta*, which transmits the doctrines taught by the Rishi Vasishta to his disciple Prince Rama. From this we learn that Dhyana (contemplation) is of three kinds: Gross, Subtle and Luminous. These are explained in detail: firstly when contemplating one's teacher (guru) or an embodiment of a god it is *stula* (gross) contemplation. When Brahma or Prakriti is contemplated as a mass of light, it is called *jyotis* (luminous) contemplation. When Brahma as a Bindu (a minute symbolic point) and the *kundalini* power are contemplated, it is called *sukshna* or subtle contemplation. Each is chosen for its appropriateness for each person and circumstance. It is valuable to be aware of these different aspects of contemplation. The contemplation of flowers would come under the heading 'gross' or natural. However, their geometry would come under the heading of 'subtle' contemplation.

'Your feet, like roots, draw power from the earth. Your body, like the trunk, is perfectly aligned. Your head is open to the heavens, like the crown of the tree. You rest calmly the universe in your mind.'
Master Lam Kam Chuen

An illuminating example of 'gross' contemplation, aiming at regaining the sense of our earthly paradise, is a veneration of the forces that lie within our natural world. This is the nurturing of the natural world and learning directly from flowers and leaves. The following Vedic commentary evokes paradise:

> Let them contemplate that there is a sea of Amrita [nectar of immortality] in their heart, that in the midst of that sea there is an island of precious stones, the very sand of which consists of pulverized diamonds and rubies.
> That on all sides of it, there are *Kadamba* trees, laden with sweet flowers, that next to these trees, like a rampart, there is a row of flowering trees and plants such as *Malati, Mallika, Jati, Kessara, Champaka, Parijata* and *Lotus* and the fragrance of these flowers is spread all around, in every quarter. [*The sevenness here is obviously cosmological.*]
> In the middle of this garden, let the contemplator [yogi] imagine that there stands a beautiful *kalpa* tree, having four branches, representing the four Vedas, and that it is full of flowers and fruits. Insects are humming there and cuckoos singing. Beneath that tree, let him imagine a rich platform of precious gems, and on that a costly throne inlaid with jewels, and that on that throne sits their particular deity, as taught to him by his teacher [guru]. Let him then contemplate the appropriate form, ornaments and vehicle of that Deity.

The Beech tree has been the place for the declaration of eternal love by generations of young people.

This method of contemplation offers the aspirant the opportunity of regaining the paradise lost by the fall — or lost on descending into incarnation. The *Kadamba* is sacred to Krishna so it is also his Arcadian Paradise. The *Kalpa* tree literally means 'time cycle tree', being axially in the centre of the garden and thereby the *axis mundi* and tree of life itself. This example shows the important mythological and psychological connection being forged between the gross and subtle worlds, tree symbolism being paramount and the vehicle.

In the Hindu tradition there is also a recommended practice of meditation sitting under five trees which were planted for that purpose. The *Panchavati* (five trees), under which Sri Ramakrishna used to meditate in the garden of Dakshinesvar, have become well known. The five trees are:

Amalaka (*Phyllantus emblica*) which is the Nelli or Gooseberry tree representing the Earth.

Udumbara (*Ficus glomerata*) a species of Fig tree representing Water.

Margossa (*Azadirachta indica*) the Neem tree representing Fire.

Vata (*Ficus bhenghalensis*) the Banyan tree representing Air.

Asvattha (*Ficus religiosa*) the Bodhi tree representing Ether.

The second, fourth and fifth trees are species of Fig.

According to this Vedic perspective all these trees exert influences on people, ranging from the physical via the intermediate to the highest. Their main qualities are strengthening, purifying, healing and inspiring. The relationship between 'material' properties and their symbolism bears witness to the intimate relationship between the various planes of existence, these planes being present in all revealed traditions.

The *Bhagavad Gita* advises us to 'resort to secluded places', particularly in a wood. Buddhist tradition has also recommended a forest life as a 'stage nine' in *Ashrama* on the way to attainment of the Highest Good.

We are be reminded of the conversation between Socrates and Phaedrus in Plato's *Phaedrus* dialogue about leaving the city for a rural setting to talk about philosophical matters. When they do eventually settle down to converse, Socrates says 'By Hera, what a lovely secluded spot! This Plane tree is very tall and flourishing, the *Agnus* is tall enough to provide excellent shade too, and since it is in full bloom it will probably make the place especially fragrant'. Without doubt a paradisal implication by Plato.

If we now turn to the Nordic tradition, we learn from the *Völuspá* about the roots of the World Tree. 'This Ash is the greatest of all trees and best: its limbs spread out over all the world and stand above heaven. Three roots of the tree uphold it and stand exceeding broad'.

The three directions or 'levels' of this tree are inhabited by three animals. The stem or trunk is the squirrel's domain and its job is to run up and down bearing the eagle's words to the serpent below. There are four deer who eat the leaves and braches of the tree, digesting the messages as food of the spirit. This is but a brief calling on what is obviously a deep and esoteric traditional teaching.

For the Christian tradition the Cross of Christianity is also symbolic of the World Tree. The European tradition of tree symbolism is similar to the Hindu inasmuch as we find the *sacred seven*. Based on his research into the ancient traditions, G. H. Mees puts the sun, the moon and the five elements into a sevenfold system (see page 84 in our chapter *The Cultural and Mythological Perspective* for a discussion of Mees's work).

Two examples of the ancient runic alphabet, one in wood and one in stone.

The Lotus is probably the flower held most sacred by the greatest number of people. From ancient Egypt to India and the far eastern realms of China and Japan, it is still held as the primary flower. The Gods of Hinduism, as well as the Lord Buddha, sit on a Lotus.

The Ash, the Mountain Ash and the Laurel represent the sphere of the sun. Ivy and Mistletoe represent the sphere of the moon, as does the Olive in Mediterranean countries (sacred to Athena/Minerva). Oaks are taken to represent the fifth sphere of *Ether* and were also considered to be 'prophets'. (Virgil testified that Oaks struck by lightning were prophesying trees.) The Maple or ", the Linden-tree and the Poplar all represent the sphere of *air*. Fruit trees, above all the Apple and the Pomegranate as well as the Larch, Yew and Pine symbolize the sphere of *fire*. The Myrtle and the Beech represent the sphere of *water*. (Note the Court of Myrtles with its pool at the Alhambra in Spain.) The sphere of the element *earth* is represented by the Cypress (sacred to Pluto as well as his underworld). In this way the sevenness of sacred trees related symbolically to the *sun*, *moon* and the five elements *earth, water, air, fire* and *ether*.

The Beech tree for the old Germanic peoples was the tree of love. A forest of Beeches was the most uplifting and 'lovely' of woods. The play of light through the leaves was especially enchanting. The smooth bark was analogous to the smooth skin of a lover. Even today lovers carve their betrothed's initials in this bark often surrounded by a heart shape — thereby pledging their love 'for ever'. The word 'write' is etymologically related to the German *reissen* meaning 'to scratch (or etch)'. 'To read' is related to the German *raten* meaning 'to guess (or advise)'. This allowed for more than a single (literal) meaning for a runic symbol.

Written records were originally scratched on Beech trees or tablets, so we find in modern German the word for 'letter' is *Buchstabe*, literally meaning 'beech stave'. The word 'book' probably derives from *bók*, the Old Norse for 'Beech'. The modern German for 'book' is *Buch* while *Buche* means 'Beech'. The written word is always in very close proximity to trees — and two pages, after all, are called a leaf.

Runes (scratched on Beech) were originally devised solely for the purpose of teaching the sacred traditions. Said to have been created by Odin the god of the Spirit, runes when used by people were taken as derivations and reflections — as distinct from those symbols used by the gods. This demonstrates the potential in levels of meaning of combined symbols. Each to his or her own understanding, the tree being ever-present in material or symbolic form in these traditions. And we still speak of a family tree.

The Bodhi tree (*Ficus religiosa*) in Bodhgaya, India. Bodhgaya is the
most important of all places to a devout Buddhist. It is where Buddha
attained enlightenment.

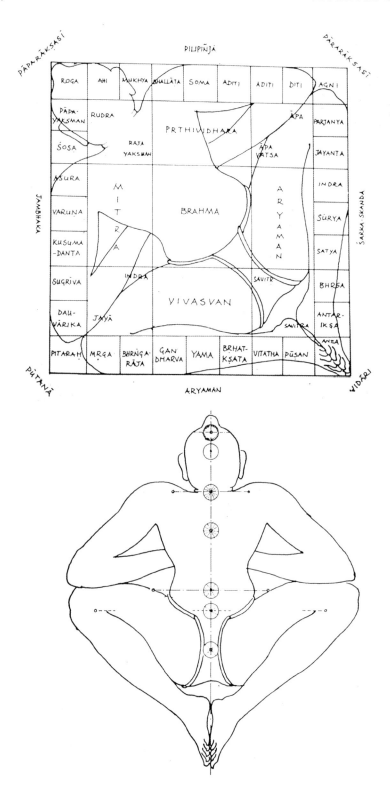

In Hindu and Buddhist understanding (and also Islamic to a lesser degree) the human body is seen as a far more subtle entity than merely its physical components. The process of healing in these cultures relies on a knowledge of the subtle channels of the body — as in acupuncture. The Vastu Purush mandala shows a symbolic image of the anthropocosm of the universe reflecting the structure of the archetypal human form (Purusha). By extracting the image of the Purusha from the mandala we can see that the positions of the lotuses or chakras were determined by the square mandala grid laid over 'him'.

Chakra organization on the body.

From Plant to Human Tree

The process of evolution, like all sequences of development, follows the pattern of the Sephirotic Tree.
Z'ev ben Shimon Halevi, from *The Kabbalist at Work*[62]

The subtle tree of our bodies is the basis for higher symbolic meaning. So we turn from natural trees to the human body with its 'lotus flower' spiritual centres in 'wheel' form. The subtle human bodily anatomy is described in detail in the Vedic and Buddhist traditions. A form of meditation is offered by the Hindu sage Gheranda suggesting this different approach to our humanness. He says: 'Let the contemplator [yogin] imagine that in the pericarp of the Great Sahashara [the thousand-petalled lotus at the crown of the head] there is a smaller lotus having twelve petals. Its colour is white, it is highly luminous, and has twelve *bija* [seed] letters, one per petal'. In this tradition 'seeds' are symbolically sacred 'sounds' as constituents of sacred mantras. The seed sounds in this smaller lotus are *'ha', 'sa', 'ksha', 'ma', 'la', 'va', 'ra', 'yum', 'ha', 'sa', 'kha',* and *'phrem'*. Also in the pericarp of this smaller lotus there are three lines forming a triangle, called *a, ka,* and *tha,* and their angles are named *ha, la and kasha*. In the centre of this triangle there is the *pranava* (the greatest word) OM (spoken as 'Aum').

Nada (meaning 'sound') represents the crescent moon, symbol of creation. *Bindu* represents the supernal sun and is the 'point'. These two are always present together at the apex of the written symbol for Om. Gheranda continues:

> 'There let them contemplate their teacher Divinity [*Guru Deva*], having two arms and two eyes and dressed in pure white, anointed with white sandal paste, *wearing garlands of white flowers*; to the left of whom stands *Sakti* of blood-red colour. By thus contemplating the teacher [guru], the gross contemplation is achieved'.

This description introduces the human body as a subtle instrument with its symbolic 'flowers' or spiritual centres in specific positions in the torso. Here we show how these are indicated by the geometry of the Vastu Purush mandala. (Vastu Purush is the anthropocosm or universe in human form. This can only be a preliminary introduction to what is a perennial subject with vast ramifications. Our intention is to introduce the doctrine of the subtle body with its lotus centres.)

Here is the image of a sound held to be the most powerful in the integrity of the universe by the Hindus and Buddhists. It is taught to be pronounced 'Aum'. It is said to be the sustaining sound of the manifest universe.

The apex of the Om sign is a symbol of both sun and moon and seals the resonance of the sacred sound.

Because of its positive characteristics (excellent wood, tasty and
nutritious nuts, medicinal leaves and shells) the Greeks and Romans
dedicated the Walnut tree to Zeus / Jupiter.

Trees and Paradise, the Wilderness and Fourness from the *Vedanta*

It is impossible to prefigure the salvation of the world in the same language by which the world has been dismembered and defaced.

Wendell Berry, from *Life Is a Miracle*

Tantra being a practical discipline ... holds that individuals contain within themselves all the essential dimensions of the universe. Tantra ideology, as is well known, works with the human model of chakra-organization.

Prof. S. K. Ramachandra Rao,
from *Sri-Chakra: Its Yantra, Mantra and Tantra*

Having established the traditions of our human nature as a subtle 'Lotus-bearing' tree, we will return to the concept of a ladder of consciousness in human nature that is of at least a fourfold order. Considering this a universal aspect to all human wisdom traditions, we now look again at the Vedanta. It is this concept of *four potential perspectives* that has been eliminated by the reductionism of today. We have proposed that flowers become a far more essential element in human life if we regard them physically, socially, culturally and inspirationally — both separately and as a whole. This is how they are viewed in both Buddhism and the Vedanta of Hinduism.

In the Vedic or Hindu tradition, we find four *Yugas* (ages), four *Ashramas* (sacred places) or four paths of life. These are related to the four branches of revealed wisdom for the four ways of life: *Brahmacharya, Grihastha, Vanaprastha* and *Sannyasa. Vanaprastha* means to dwell in the forest, to nourish sacrificial fires. The forest is a symbol of the mental plane — 'the wilderness' — that has to be brought under control by meditative discipline. The mind is a wilderness until it is harnessed. Each of these four perspectives suits the natural divisions in human nature — according to Hindu tradition and to individual appropiateness.

For the Vedic tradition there are four ends of life: the *Purushartha*s or 'Human Ends'. These can be summarized in the following chart:

A tree in London's Green Park.

States of Being in Sanskrit	Ends of Life (purpose)
4. *Moksha*	Spiritual element, salvation and *liberation*
3. *Dharma*	Social fitness, reasonableness, *rationality*
2. *Kama*	Emotional life, desires, *social interrelatedness*
1. *Artha*	Wealth and health, *bodily comfort*

The above four can also be seen as a ladder of both awareness and individual development and responsibility.

There are also four *yogas* or disciplines, four *padas* or paths, four *margas* or ways. These words in Sanskrit are far deeper and wider in meaning than the English terms. Each is suited to a temperament and what is now superficially called a 'life-style' in the West.

Commenting on the above, G. H. Mees writes:

> By uniting the fragments in their own life and soul to a living body, a person unites the fragments of the world around them and transforms them into the 'heavenly city'. Engaged in that one pointed end, they will use more and more of their self-consciousness and become ever more loving and serviceable to others; not like a machine turning out a manufactured article, but like a flower opening up its heart to the sun.[63]

We can symbolically read the 'True Self' or the 'Kingdom of Heaven' as the *Satyaloka* or the paradise within the heart self and the source and goal of all life. 'Paradise', as Mees reminds us, is 'the most real state of all'; that is, the *unchanging* reality of the intelligible realm. This deeper meaning of 'paradise' is not fundamentally different from Socrates' injunction that the 'Soul' is immortal. This places our relationship with the floral world within an aspiration to regain our original paradisal state and regain the wholeness of our soul.

To complete this part of the book we will turn to the word 'universe' which is itself a reflection of paradise. 'Universe' comes from the Latin *unus* ('one') and *vertere* ('turn'). It is the ever-turning 'one' or, as the *Oxford Dictionary of English Etymology* has it, 'turned into one' — reminding us of the consistent turning of the petals as a flower emerges from its bud. Boethius clarifies this turning beautifully in his *Consolations* (II, viii, 13):

> What binds all things to order governing
> earth and sea and sky is love.
> If love's rein slackened all things
> now held by mutual love
> At once would fall to warring with each other.

Socrates speaks many times of the role of geometry in turning the soul inward to its original, remembered state. In Plato's *Republic* dialogue he calls this 'metastrophe'. Flowers naturally 'turn' our thoughts to Paradise.

The Eye of the Heart

Such is the traffic of perception:
these mutable skies, reiterative seas,
waxing and waning of featured land
held in a nature not their own.
Each sense cognizes a domain:
desire's pulse enflames the red rose,
serenity embraces blue,
the green world hoards
and prospers our repose.
Each reaches out to colour
what seems its own
yet takes upon itself to err.

What dawn invokes, night betrays,
fading and dying to the dark.

So let these eyes be closed
though love most readily enters there.
Let seeing be such as
reaches into the heart
where every sense, assembled,
knows as one
illumination is by light alone.

Brian Keeble

Kukai, Buddhist master of ancient Japan.
Shown in his timeless spherical body sitting on the sacred Lotus flower.
Note the three Lotus seeds on the head of the seedpod on which he sits.

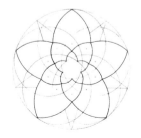

SIX

CONCLUSIONS

No net will hold it — always it will return
When the ripples settle, and the sand —
It lives unmoved, equated with the stream,
As flowers are fit for air, man for his dream.

Kathleen Raine, from *In The Beck*

The natural habitat or planetary setting for us as humans consists of light,
mountains, seas, rivers, trees, flowers and creatures.

So, What Is Life That Flowers Represent So Magnificently?

Where is the Life we have lost in living?
Where is the wisdom we have lost in knowledge?
Where is the knowledge we have lost in information?

T. S. Eliot, from *The Rock*

For those who have forgotten, or those who are convinced otherwise, we need to restate our own conviction that life *is* a miracle. If some are not sure what a miracle is, then at least we have the Oxford English Dictionary to help in this matter. A miracle is 'a marvellous event occurring within human experience, which cannot have been brought about by human power or by the operation of any natural agency, and must therefore be ascribed to the special intervention of the Deity or of some supernatural being'. ('Supernatural' literally means 'above' the natural or even prior to the natural. The 'natural' means the nature of the 'world' around us — named by the Greeks the 'world of generation or change' and therefore only a reflection of its intelligible source.) The word 'miracle' originates in the Latin *miraculum*, related to *mirare* meaning 'to wonder at'. *As there is no current definition of 'life' from within the empirical scientific community*, only of what 'life' does, we propose that it is quite acceptable within normal usage to call 'life' miraculous *and propose that this is also a scientific definition.*[64]

We have all been introduced to the concept of entropy, which has been directly associated with the cessation of thermal energy for the conversion into mechanical work. However, upon further investigation we find that the definition of 'energy', from the viewpoint of the physicist, is merely 'a body's power of doing work by virtue of its motion'. The word 'energy' derives from the ancient Greek word *energeia*, which comes from *ergon* meaning 'work'. This led the Western scientific community in the 1970s to announce their *extraordinary* conclusion that, 'The only ascertainable purpose for life is to delay the dissipation of solar energy' (an absurdity which seems comparable to looking through a telescope from the wrong end). This pronouncement demonstrates how, by taking a view defined by extreme narrowness, one can come to a conclusion that defies even common sense.

Rupert Sheldrake, the distinguished British scientist, has kindly

reminded the author of the two words that have generated our modern English terms in relation to life, namely the Greek *zoe* and *bios*. *Zoe* is the ancient Greek for 'life', 'to live or exist'. *Bios* refers to the *living* of one's life, the passing through of one's life. We are all familiar with *biography* (the story of a person's life) or *zoology* (the science or study of animal life) but how many of us have asked ourselves, 'What is life itself?' Yet we all experience in one way or another what living is while we are alive. We are all aware that life is never straightforward or automatic due to our being able, and desiring, to make decisions based on freewill. We all have an idea of what 'making a living' means, probably reducing man's life to a basic necessity as it takes into consideration only the financial or material dimension. However, due to the current, nearly tyrannical, domination of the advertizing of the commercial world and the news and entertainment media, such a definition is virtually all-pervasive in its influence. The author cannot but enjoy the cryptic directive 'get a life', with all the meanings that could be attached to it. It surely is exactly what each flowering plant sets out to do.

Dr Martin Lings, another contemporary luminary, said: 'The mark of true humanity is the ability to receive inspired insights'. This is the nature of those who have changed whole aspects of civilization, and even the very civilizations themselves. It is often proposed that these inspired insights come from a given 'poetic vision' or through intuition. ('*Int*uition' is another word that has suffered from recent debasement as it means literally being taught — given tuition — from within. This could be offered as another dimension of Platonic remembering or awareness.) Such a perspective is unfolded poetically by the inspired Islamic poet, Jeláluddin Rumi, in relation to the greater dimension of meaning of the word 'evolution'.

> I died as stone and became plant,
> And from vegetable I died to become animal.
> I died as animal and became human.
> When was I less by dying?
> Next time I shall die
> And become winged like angels —
> Beyond that soaring higher than angels
> To what is unimaginable.
> I shall become that.[65]

This does not allude to a physical evolution which is taking place over thousands or millions of years, but to states of being in transformation, happening in the 'now' of time. In the great traditions, everything is evolving and devolving in this world of change. For the present author, modern theories of physical evolution are tragic parodies of fundamental metaphysical truth.

From here we return to our opening observations that a careful distillation of traditional teachings would seem to indicate that there are at least four levels of being. The lowest is the 'apparent' illusive state of physicality and constant change. The next level that transcends this is the psychological level of societal awareness. This is the kinship of all human relations. Above that we have the created treasures kept by the establishment of culture, poetic, imaginative, sculptural, painterly, musical, literary and even mathematical. Finally, interpenetrating all of these, and moving beyond the level of culture we have the inspirational, recorded in the timeless revelations of Divine Beings. It is upon these that civilizations are founded, maintained or dissolved. This final level is the motivating energy of all the rest. Each 'level' invites participants 'up' to a more subtle state of understanding and being. Each level offers its own panorama of vision, and each offers its aspects of 'permanence' according to the ability of the receiver to accept what is proffered.

The mystery of life that runs through all things that we recognize as 'living' is not diminished through our having discovered so much about its exquisite ramifications and details. As we add to our knowledge, we see that life is more miraculous rather than less. According to visionary US biologist Lynn Margulis: 'Life did not take over the globe by combat, but by networking' (in other words by cooperation and collaboration, as we have already noted).[66] In contrast, too often the attitude and viewpoint of today is to ascribe an isolated 'individuality' to any living thing. What has come to be called 'the selfish gene' is a prime example of this almost unbelievably narrow and fractionalist view. Naturally all life and living things have a survival mechanism built in, but *collective survival* is more revealing and far more pertinent. What immediately comes to mind is the way shoals of fish collectively move to avoid the much larger predator or the mysterious way in which a flight of starlings moves simultaneously like a dancing cloud, with no known explanation.

Next we note the tribal instinct, or the synergy of human societies, which is founded on inherent social laws which harmonize the group concerned and increase its ability to survive. Beyond this we might cite the way plants group or even complement other species by collaborating in their distribution of minerals, pollens, light-taking and so on. As in all

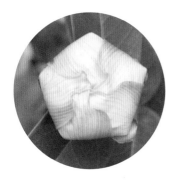

studies, the most important discipline is to attempt to see the largest picture *as well as* the detail. Often the whole can be found and understood within the part if it is looked for in the right way.

What is more, the surface of the earth, which we have always considered to be the environment of life, is really *part* of life, in fact a product of life. When scientists tell us that life 'adapts' to an essentially dead environment of chemistry, physics and rocks, they perpetuate a severely distorted view. Life actually moves and forms and changes the environment to which it has 'adapted': most surface rocks on earth were created by life. These conclusions are far more in keeping with the ancient Chinese concept of the Tao than with the many partial contemporary attempts to restrict life to mere description. For Plato the universe was a living being with an intangible soul that sustained it. The world was 'of the changing', the 'soul immortal' or 'a permanent principle'. After years attempting to reduce all of our experience of life and the environment to physical actions and reactions, we have inevitably reached the point of nihilistic crisis where we can no longer see the wood for the trees, or the whole for its detailed parts. The sages of past time speak of an intangible harmony that sustained all existence. 'If it is not of the Tao [way] then it will not last', as the classic of Taoism, the *Tao Te Ching*, states. This posits a universal order of which we can become aware through stillness, silence and our own inner tuition, our *in*tuition. In this context, we acknowledge again the important work of Rupert Sheldrake, another courageous scientist who has had the insight to embrace the time-tested techniques of ancient Vedic meditation so that he (the responsible instrument of his own ideas) is integral with the emergent ideas themselves. The concepts that Sheldrake works with, such as that of 'morphogenetic fields', seem to us to be leading in the right direction.[67] That is, concepts which approach our universe and our selves in a more wholesome and healthy way as a story of the 'fittest to collaborate'.

Flowers perpetually remind us that at the top of our list will inevitably be the mystery of life itself. We are immensely clever and erudite in telling what life *does* and how it *constructs* but completely inadequate in saying what life *is*. We irresistibly manipulate living organisms predominantly for our own narrow benefit. We are losing the ancient wisdom of the wise farmer who works *with* the natural world. We are thankfully extremely good at 'mending' living things when they are in ill-health. But this still does not bring us to a definition of life itself. Primarily we have to *live* it.

This is where a layered hierarchical model is so valuable, as we can be a little more sure *from which perspective* we are regarding the question. If we return to the 'ladder' of perception[68] we used before, we can frame the question of what is life in quite different ways — each of which will offer a different perspective. The 'Divided Line' of Socrates when correctly read (vertically) in Plato's *Republic* also reveals the four-layered tradition of the powers of knowing available to us.[69]

There are pointers that even common sense indicates:

a) Materialism by itself inevitably leads to nihilism. Life is not matter, it *raises* matter.

b) From the social perspective quite a new series of possibilities arises. Life *is*, it is lived, it does not require an intellectual definition as it lies in the very living relationships between all things, especially between human and human. It does not require a 'definition' to produce another human being! It requires only participating in the miracle and the source energies — love and trust.

c) The cultural standpoint is developed out of the inspiring and sustaining ideas on which life's laws are framed. Life succeeds by being coded into ethics or beneficial behaviours between peoples. Possibly the most universal example of this is 'Do unto others as you would have them to do unto you'. Each culture develops its ethos from the 'highest' sources available to it. Life is dependent on what culture there is to nourish and nurture it as harmoniously as possible. Civilization is the fruit of this harmony.

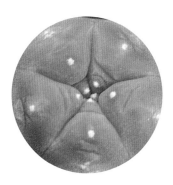

d) The inspirational level clearly emphasizes that the definition of life as a rational simply-understood concept is not available to human consciousness as such, and can only be described as the gift from the highest Good. This could be the reason why the Lord Buddha would not answer any questions on either the existence or non-existence of God. The discussions, he knew, would only encourage division and probable conflict. Yet he did not reject the word 'Brahman'.

Realization, as in all authentic traditions, reaches beyond mere words and definitions into direct *ineffable* experience. This brings us back to flowers which were used by the Lord Buddha to express the ineffable. On one occasion when a sermon was expected by a large number of his devotees *he simply held up a flower and spoke no words that day*. Some of those present entered enlightenment at this event, as the reports go.

When the presence of Beauty as a perfect harmony is appreciated, the need for explanation or definition is cancelled immediately. Both the viewer and the viewed become 'one' in the viewing. This however could be seen as deeply disappointing to those who live by contention, competition and constant debate.

The mysterious relationship between the honey-gathering bee and the nectar-
giving plant world is not a simple mechanical one. Bees communicate the source
of their findings by dancing in the hive. Different species of bee may show a
preference for different flower geometries.

A Flower and its Existence

I know a bank where the wild thyme blows,
Where oxlips and the nodding violet grows,
Quite over-canopied with luscious woodbine,
With sweet musk-roses, and with eglantine.

Shakespeare, from *A Midsummer Night's Dream*

The relationship between ourselves and flowers and plants will always remain essentially powerful yet mysterious. As we noted earlier, Albert Einstein is said to have frequently quoted the Biblical phrase, 'All flesh is grass'. We need particularly at this point in human history, where so many and so much of our foodstuffs are artificially processed, canned and packaged, to be reminded of the total debt that we owe to the vegetal world for our own bodily fabric, just as we need to be reminded of the essential relationship between the insect world — particularly the bees — and the fertility of plants. No bees, no plants. No plants, no people. No pollen, no flowers. If there is no pollination, there can be no new generation of the vegetal kingdom. The bee is so important as a bridge between the worlds of plant and insect and human that there is even a whole chapter (*Surah*) of the The Holy Koran entitled 'The Bee'. (In the Judaeo-Christian *Old Testament* bees are famously mentioned in Judges 14:8, where Samson is fed by honey in the body of the lion he has killed. From which we get the riddle: 'Out of the strong came forth sweetness'.) It is essential that there is a continuous transfer of vital energies from the mineral world of the soils transformed through the vegetal world of plants, and then through to the animal world and right up to the consciousness of the human world. Our recovery of a sense of ecological — particularly spiritual or metaphysical — wholeness is essential for our sustenance.

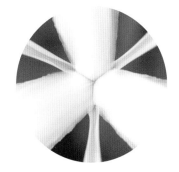

If we can study life as a whole, that is the totality of the levels of energy transfer and the way each aspect of life harnesses that which is necessary for its continuance, then we may be nearer to understanding what attitude we should take towards each part of this whole and what our role in the whole scheme of things should be. Hopefully, this discipline will also enable us to recover from the imbalances we have created in the contemporary world. All of us today are aware of the urgency of our circumstances and the sacrifices we have made through industrialization.

It has occurred to me, having taught in British universities for half a century, that maybe all heads of department in all institutes of higher education should, as a duty, give an opening speech each year on behalf of their department or school explaining in simple terms how their particular subject benefits humankind and life as a whole.

If we, as flesh and bone, are made up of the nourishment that originates in the vegetal kingdom, then we have even further reason for recalling our indebtedness. For the present author, the visionary words of the poet are most likely to awaken a higher reality in us, a reality more permanent and instructive than that created by a logical mind locked into its own conventions. It is worth recalling Jeláluddin Rumi's question in *The Mathnavi*:

> The spiritual inspiration of man's knowledge bears him
> aloft; the sensual empirical basis of man's knowledge is
> a burden. Do you know any name without a reality? Or
> have you plucked a rose from R-O-S-E?

Prior to the current tyranny of the materialistic, mechanical model of our universe, the wise people of all traditional cultures viewed the world as a continuous spectrum. This spectrum spanned from the pre-consciousness of absolute subtlety to the most gross expression of changing materiality — the subtle world being that of permanency. Again, as Rumi writes in *The Mathnavi*:

> If thou wilt be observant and vigilant, thou wilt see at
> every moment the response to thy action. Be observant if
> thou wouldst have a pure heart, for something is born to
> thee in consequence of every action.

Here we learn of the traditional doctrine of the continuity of consciousness and the concept of the great Chain of Being that links everything to everything else in a continuum. From this perspective, it means we are placed not only in a necessary relationship with the plant world in the continuity of physical existence, but also within the spectrum of consciousness by which we can 'know' anything.

What is becoming increasingly urgent, as we have indicated, is that the modernists of today's world need to change their perception of the totality of things from competition to one which is based on complete interdependency; and to see each of us as responsible, not only for our own wholeness, but also for the whole of the human family and the natural world as well. Any other path will increase conflict and deprive us of our collective birthright, that of peace of mind and body. We can only 'flower' ourselves when we pay due reverence to the flowering plant world and the primary elements that keep them in being. In this book we have presented the evidence embodied in the flowers and leaves which collaborate with human intelligence — a collaboration having more do to with the growth of intelligence in our evolution than with the implied conflict of 'survival of the fittest'.

As nourishment is the key to all our levels of awareness we inevitably have to arrive at light as being primary. Light illuminates at every level, inspirationally (spiritually), intellectually, traditionally, culturally, socially, and obviously botanically and physically. Photosynthesis is a word we use about plant building but it is also the most powerful metaphor for the seeking of wisdom itself. What else is enlightenment?

Apparently, it has taken millions of years to establish the best working relationships between all the diverse creatures and their physical settings, particularly to arrive at what we call 'collective civilized behaviour' between people. Surely it is time to take stock of what is vital to preserve and what is vital to avoid in our immediate living patterns. Our gift of self-determination or 'free will' is a huge responsibility and never more so than in our own present times. What *do* we want or need?

So much depends on how we take stock. This means how careful we are in regarding how and 'from where' we take our 'view point' or our 'stance' and deeply consider what we *really* need. In his poem *Below* (1980), the Kentucky writer and farmer Wendell Berry concludes:

> What I stand for
> is what I stand on.

Our admitted ignorance of how the 'whole' works might and should encourage a little more modesty and humbleness. It is in just this area of perception (and depth of understanding of what perception means) that we are in most need of care and attention. Continuing death-reliant analysis and genetic engineering do not recommend themselves.

It has been said in many ways throughout the time of human development that we are constantly engaged in the 'flowering of our consciousness'. This means we are perpetually prompted to ask *who we are, why we are here, how we came to be here, what we should be doing* and so on. These questions of personal meaning have given rise to many different solutions but with recognizably common symbolic meanings. A scale of fourness has arisen time and again in human wisdom traditions from China to India, from indigenous America to Judaism, Christianity and Islam. They can be taken as the steps to wisdom.

In Summary

The truth is that life is not material and that the life stream is not a substance.

Luther Burbank

Let us return to the word 'nature'. What we can be certain of is that we are not, and never have been, separate from nature while in human form. The Latin word *natura* means 'birth' as well as 'the course of things', so 'nature' refers to all things that come into existence, live and develop, mature, age and die. However, it also means 'essential qualities' or innate character. This touches on the deeper meaning of principles that lie within all that lives. We also find in the *Oxford Dictionary of English Etymology* another more evocative meaning: 'creative and regulative power in the world'.

Having taken you, the reader, through a systematic analysis of the form in flowers, in particular the geometric order intrinsic to all flowers, now is the time to make as simple a summary as possible.

As 'nature' is simple as an embracing principle and experience, so is geometry. Most who have been exposed to geometry at school probably enjoyed the drawing aspect, as the author did, but found the emphasis on 'proof' a rather taxing abstraction. The truth is that geometry is essentially a simple art based on the natural order of space. The need for space, for time, for number and form we all acknowledge. They are summarized in 'substance' or that sense of materiality that 'stands under' all physical experience.

We find all things that are natural come into being at a given point. This can be called the 'point of departure'. In the plant world, this point of departure is also the point of origin and known as the 'seed'. In the language of geometry this point, in order to develop, proceeds to create a movement which creates a 'line'. This is 'direction' and for the plant it is a shoot. Euclid calls the line the *first* dimension. The line, by moving in any other direction, will create or leave the trace of a surface or 'plane'. This is called the *second* dimension. The moves are simple and head towards generating a 'solid' space. The plant world expresses the plane or surface as the leaf. Any further movement by the plane which is not

Those who live closely with the natural world can read the sky, the trees and the flowers. We are inseparable from our setting as human beings. In the photo is Thomas Banyacya, silversmith, jeweller and sculptor, representative of the Hopi people.

There are still human communities that harvest the grass-of-the-sea or seaweed as an excellent nourishing source for people, animals and the land.

The Lotus in the natural world. Here, a 'field' or lake in China with a multitude of flowering plants. This demonstrates how their flowers rise above the water surface, which the Water Lilies (*Nymphaea*) do not.

For thousands of years of human cultivation, Poppies and the blue Cornflower all grew together with the corn. Today they are treated as 'weeds' and exterminated as far as agribusiness can. Commercial criteria are unlikely to accord with ecological or even aesthetic criteria.

in the direction in which it was created will trace out a solid space: the *third* dimension.[70]

Each of these factors has its own mysterious quality, not least the *meaning* of 'dimension' itself. A word easily accepted by us all yet its definition is not so easy. A clue can be found in the 'di' of *d*imension which means 'two' (we usually measure from a point to a point) but still where does this get us? Dimension is most commonly understood as meaning measurable. However, if we come back to the spatial simplicity of gaining dimensions to arrive at the world we experience (which has by necessity *three* dimensions) we have still had to overlook the 'point'. Why? Because Euclid tells us that the point has 'no part', so it has 'no dimension'. It is only by its movement that we get any dimension traced.

Thus the significance of the point, the veritable starting place of all spatial order, is immeasurable. Surely the following question must arise. Does the point exist? If so, what are its qualities and what can we know of it? Euclid in the great traditions of Plato and his master Pythagoras offers evidence that it is a metaphysical reality, not a physical actuality. Not only is the point a metaphysical concept, but so equally are the lines and the surface or plane, as it is not until we enter the *third dimension* that we are describing an 'actuality'. In short, physical existence takes place in three dimensions. Prior to this we are in the domain of concepts or ideas. However, a flower arises quite naturally through the dimensions, having started at the 'point' seed.

This is not to deny the immense power of the 'point' as 'seed' in the biological domain, but rather it emphasizes the intrinsic power of the idea of a point. (We have already considered how modern astrophysicists speak of a 'point singularity' as the origin of their Big Bang theory.) This brings us back to the critical importance of flower forms in the evolution of human intelligence.

Albert Einstein placed wonderment at the apex of human qualities, 'without which', he said, 'we are as dead people'. What better companions in our shared planetary life than the flowers at which we wonder?

Life and death are the basis of the drama we call living, yet life so far has been continuous since its advent. As there are four seasons experienced by the Northern civilizations and as there are four 'elements', it was soon established that a principle of 'fourness' pervaded our natural condition in many ways. If we consider the directions (north, south, east and west) as well as the periods of human life (childhood, youth, maturity and old age) it is not surprising that a universal concept of fourness emerged with regard to senses, mind, knowledge and understanding. This ranged from estimations to opinions, then from knowledge to wisdom or full understanding. This offered humanity a ladder of sanity, clarity and realization: how to view the world healthily, without violence, without conflict, but with affection and above all with compassion, purpose and a love of life. This brings us back to our beautiful fellow-travellers in life — the flowers.

We have already noted that the Lotus is probably the most venerated of single flowers in the history of human civilizations. Their worship, if it may be characterized as such, spans the early kingdoms of Egypt to the spread of both Hinduism and Buddhism in the East. Here we have a chance to look at the role of the Lotus in all four dimensions in our thesis of levels. The Lotus (*Nelumbo nucifera*) is not only a magnificently beautiful bloom but it displays its 'point seeds' most significantly and obviously at the centre of its petals and on a type of platform top to its seedpod.

We have made close relationships with Roses for as long as there are records. They bring joy, they bring sublime perfume, they bear exquisite colours and are the field of nourishment for insects which in turn fertilize them.

I know of no other flower that displays its seeds in such a manner. Plato suggests humankind learned its point-numbering or arithmetic from looking at the night sky. I would be most surprised if some now unknown Buddhist sage had not suggested equally that point-numbering arose from counting the seeds in the pod-head of the Lotus.

In a similar way that the light-points in the night sky were believed to be the home of the gods for the ancient Greeks, the Lotus seedpod-head became the 'throne' for all future representations of the Lord Buddha. In ancient Egyptian mythology vital elements — even the gods — themselves were 'born' out of the Lotus. For instance Horus the Egyptian god of stillness and silence, child of Osiris and Isis, was cradled in a Lotus leaf.

In the Buddhist tradition the Lotus is said to represent 'self-creation', the 'self' being one's Buddha-nature. In ancient Egypt, Isis herself was represented as a Lotus bud, having arisen from the water.

Horus her son, the hawk-headed one, was described as 'the one that is born with the sun each morning'. There is a remarkable parallel between the images of ancient Egypt and those of much later Buddhism. In the Hokke-ji Temple of the Lotus Sutra in Nara, Japan, we find a representation of the deity Juichimen Kannon surrounded by her halo or subtle body which is composed of Lotus stalks, bud, leaves and flowers. An almost identical surround for Horus is found in ancient Egyptian iconography. This bird-like god is also surrounded by a radiant subtle body composed of Lotus stalks, buds, flowers, and leaves (see illustration on the following page). The resemblance between the two is striking and surely contains a mystery yet to be explained.

Chartres: Great South Rose Window.

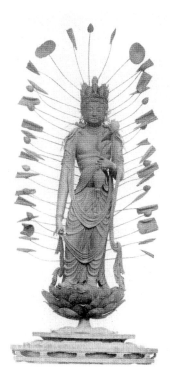

The Lotus subtle body of Kannon or Kuan Yin, Goddess of Mercy. See the written notes around this image.

Sengai's universe image was only made of a triangle a rectangle and a circle. Here there are obvious, yet mysterious references to these primary shapes. This well known yet profound simplification of our known world compressed into three shapes. The first is the circle on the right which becomes the triangle in the centre again in turn becoming the rectangle. This is the Master Zen artist's version of the universe.

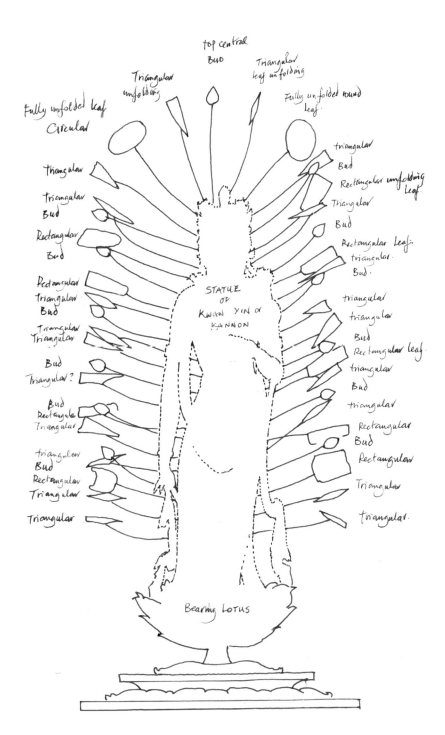

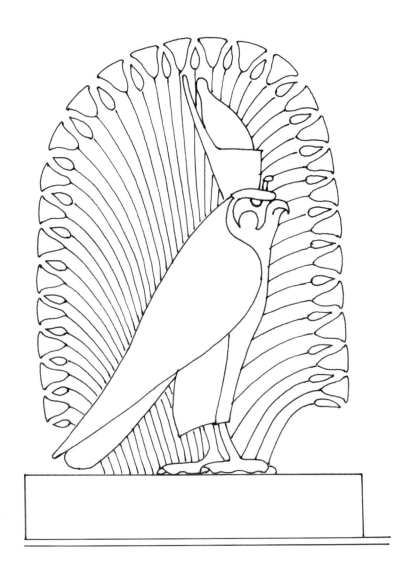

The Willow catkin as subtle body.

Here, we show a remarkable parallel between the surrounding 'subtle body' of Horus (the Egyptian hawk god) and the subtle body of the standing statue of Kannon (Kuan Yin) in the Temple of Hokke-ji in Japan (opposite page). Both 'auras' or subtle bodies are composed of Lotus leaves, buds and flowers, as we can see! The numbers are most likely to be of special significance.

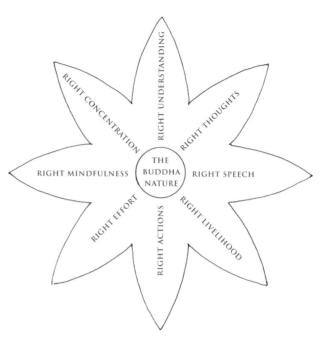

This flower has served the highest aspirations of human societies over thousands of years from ancient Egypt to contemporary China, India, Tibet and Japan.

Another important aspect of the Lotus is its adherence to the number eight in the symmetry of its petals. To relate this eightness to the important mythological number of ancient Egypt, one would have to include the vital central ninth element of the seedpod. In ancient Egyptian theology their nine gods were presented as an ennead. Nine was their most revered number. For Buddhists, one cannot but reflect on the essential noble eightfold path: the eight petals of the Lotus each representing one of eight principles and the centre being taken by the unifying Lord Buddha himself. He is invariably seated on the seedpod with its potent arithmetic point-seeds. This could be taken as symbolic of the profound unity of enlightenment surmounting multiplicity. Each petal represents of one of the essential eight paths, thus:

1. Right understanding — free from superstition and delusion
2. Right thoughts — elevated and worthy of intelligence
3. Right speech — kindly, open and truthful
4. Right livelihood — bringing hurt or danger to no living being
5. Right actions — peaceful, honest and pure
6. Right effort — in self-training and in self-control
7. Right mindfulness — the active watchful mind
8. Right concentration — deeply meditating the realities of life

The eight paths are eternally valid for all Buddhists. The commentaries vary and those given here are contemporary.

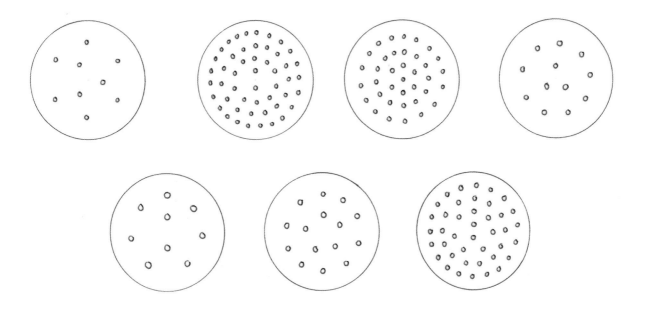

Firstly, the seedpod of each lotus has an extraordinary apparent freedom of choice of numbers — we show a few examples. These do not follow the economic laws of close-packing. But they are each number patterns that challenge the human mind by their very differences.

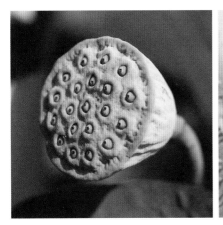 

The birth of point-numbering.

On the seedpod head of a Lotus flower the seeds of the next generation are displayed in points in many different patterns.

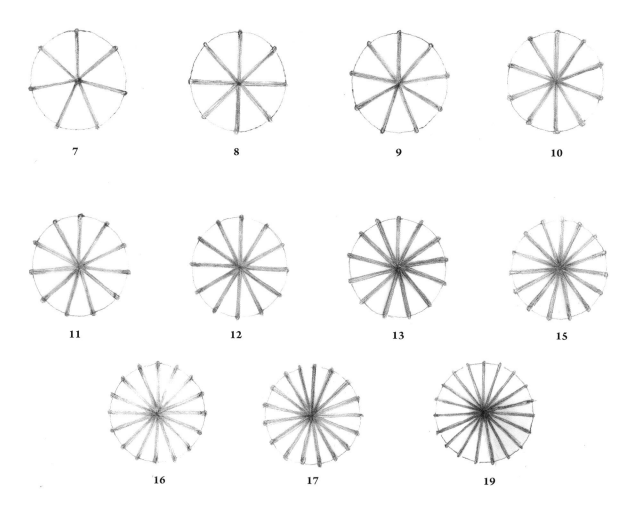

The Poppy seedpod holds thousands of minute seeds in its 'body'. However, on the top of each Poppy pod is a series of what can correctly be called radii and diameters. These eventually enumerate the 'windows' that open to let out its multiple seeds. We show a selection of Poppy seedpod heads to demonstrate their equally challenging geometry to the human mind.

Left: This Poppy demonstrates the visual power of fourfold symmetry yet has a ninefold seedhead.

Right: Again we have a powerful four-petalled poppy, but this time with a tenfold seedpod symmetry — why?

Having found inspired meaning in the Lotus for both the ancient Egyptians and the contemporary Buddhists, we will now turn to the geometrical significance of the Poppy (*Papaver somniferum*). As we carefully noted earlier, the point being a purely metaphysical principle is described as 'moving' in a particular direction to trace a line. As profound as the point-seed is in the Lotus head, we now have the next most important aspect of geometry exhibited by the radii and diameter lines on the Poppy seedpod head. A radius 'radiates' from the centre point to define the size of the outer circle. The diameter measures across the circle and is exactly equal to two radii.

Although the Poppy does not hold anything like the spiritual or mythological power of the Lotus, it does have an important role in its medicinal value — quite apart from its extraordinary linear geometric patterning on its central seedhead (see Poppy illustrations).

There appears to be no simple logic to the variations in the radii and diameters of the Poppy head compared with the logic of the point and seed patterning of the Lotus head. The way in which these beautiful central features of the Poppy flower display their patterns is quite mysterious and wondrous. If the Lotus seedpod head is an invitation to human curiosity about point numbers, so the Poppy seedpod head is an invitation to discover the significance of the circle-defining radii and diameter — as geometrically they only have *one* dimension.

As challenging as these patterns are in the Poppy, there seems to be no clear reason for the mystery of the number of lines, which appears to be quite independent of the number of petals of the Poppy flower itself. The Poppy was sacred to Demeter, the Greek goddess of the green world, associated with the moon and a symbol of fertility. It was used in the Eleusinian Mysteries, and was a symbol of consolation. The gods Hypnos (the god of sleep), " (torch-bearing death) and Nyx (goddess of night) all wear wreaths of Poppy capsules and hold Poppy flowers. The Egyptians also used Poppies for wreaths. The heads of Ceres, Libertas and Bonus Eventus (good outcome) were wreathed with Poppy flowers as was the sleeping Eros. Alexander the Great of Macedonia is especially associated with the Poppy having brought opium to both Persia and India. Thanks to the trade routes of the Arabs, the Greeks and the Romans all used it as a sedative. In our present context we wish to lay the emphasis on its geometrically challenging seedhead.

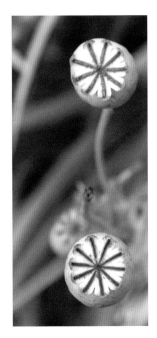

Poppy seedheads developing after losing the petals. Here we see how the 'flat' seedhead of a Poppy begins to contract upward and flattens out. This makes the line geometry even more insistant as well as beginning to open the 'windows' in the pod below to release the minute seeds.

Here we have a mystery equal to the seed patterns of the Lotus on its podhead. There is a mathematical pattern here but with no obvious 'law' or logic, certainly a prompt for the human imagination.

In this Crysanthemum we have a beautiful example of spherical spiral forms. Note how the Crysanthemum brings up its spiral petals from a central inner vortex — yellow that turns white as it spirals outward.

The fir cone also demonstrates its right and left spiral nature.

The sages of earlier human societies taught that 'nature' was the great educator. The Chinese Master Chu Tzu or Chu Hsi (AD 1130–1200) says in his commentary on Yang and Yin, the two most fundamental universal forces:

> The teachings concerning the Yang and Yin and the five elements (or agents of materiality) are made visible in the book of Nature so that humanity may learn.

It was the same master who affirmed that 'the man of humanity regards Heaven and Earth and all things as one body', thereby upholding the first principle of integrality as did Socrates. The upper world became associated with 'perfection' and 'permanence', and the lower 'material' world was understood to be always changing and impermanent. We have offered our findings on flowers as a 'bridge' between the 'upper' and 'lower' worlds. The Lotus which roots in the mud and slime climbs through the waters to bloom in the airy sunlight. This became a perfect symbol for the development from virtual chaos to sublime light, or from ignorance to enlightenment.

It is not surprising that flowers have also been so instrumental in forming human ideas of paradise. We have already quoted G. H. Mees: 'paradise is the most real state of all', because it is the most unchanging reality of the intelligible realm. Experiencing the plant world and in particular the floral aspect of this 'heavenly' domain, we are all to some degree 'reminded' (through inspiration and intuition) that these exquisite things called flowers come from another more perfect realm. This leads us to what could be called the ultimate climax or purpose and joy of flowers — learning of paradise or the opportunity of enlightenment itself, as an all-embracing fulfilment of life.

As it is light that brings life-energy to our world so it is light as a multivalent symbol that brings our minds and imaginations to paradise. Light obviously works on all four levels of our suggested ladder of consciousness. As we have seen, *photosynthesis* is a perfect metaphor for all four levels. In the sensible world light is the force that draws a seedling up from the soil into the air and then enables it to grow (just as it enables us to perceive this). Light synthesized in the leaf becomes a metaphor for light synthesized by the mind to gain understanding. Finally, enlightenment or the experience of *being* 'light' is offered as a goal for those who have given their lives to spiritual attainment.

There are two distinct paths or orthodoxies that we can refer to for confirmation of this. The first is in the teachings of the Lord Buddha, the second is in the powerful poetic vision of Dante in his *The Divine Comedy*. Our previously cited authority on parallel traditional symbolism, G. H. Mees, says that there is a direct correspondence between the Lotus and the Rose and the Golden Flower of the Chinese tradition. Mees also draws attention to the connection between the Lotus and the fifth element: ether. This element is also called *Prajna* in the Vedic perspective, indicating 'understanding', 'insight', 'wisdom' and 'knowledge' — all properties of the mind in the 'intelligible' realm of the Socratic 'Divided Line' (see page 52).

On this aspect of symbolism and the value of mythology we have the excellent explanation given by Iamblichus (AD 250 – c. 325) in his *Theurgia or The Egyptian Mysteries*:

The archetype for Dante's vision of Paradise. The white Rose holds a sublime range of meaning and value to us as human beings.

> God's being shown seated on a Lotus signifies a
> superiority which arises above and excludes all contact
> with the mind of the world. It also points to the reign of
> the intellect in the heavens. For every feature of the Lotus
> is circular, from the outline of its leaves to the shape of
> its fruit, and circularity alone is akin to the activity of the
> intellect as it invariably manifests itself in identity, ruled
> by one order and one reason [the central point]. God
> himself is established in himself as being above power and
> activity of this kind, august and holy in his transcendental
> simplicity, abiding within himself — this is what this
> being seated on the Lotus signifies.

And finally we come back to Dante's *The Divine Comedy*,[71] in particular to the climax of his spiritual journey to the Rose of Paradise. Dante wrote:

> In that heaven which partakes most of His light
>   I have been, and have beheld such things as who
>   Comes down thence has no wit nor power to write;

He explains why the human mind finds it so difficult:

> Such depth our understanding deepens to
>   When it draws near unto its longing's home
>   That memory cannot backward with it go.

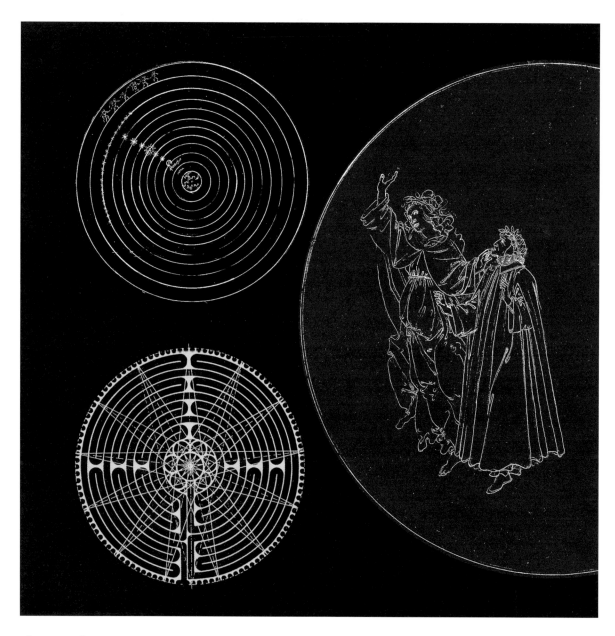

Above: *Paradiso* II
First planetary sphere (Heaven of the Moon). Beatrice explains to Dante the origin of the dark patches
on the moon's surface, then the order of the cosmos.
Image inset: (bottom left) The Chartres labyrinth suggests a shared cosmology.

Opposite above: *Paradiso* VII
Second planetary sphere (Heaven of Mercury). Beatrice introduces Dante to a number of complex issues
concerning Divine Justice.

Opposite below: *Paradiso* XXIV
Eighth sphere of Heaven (The Fixed Stars). Beatrice entreats St Peter to examine the true faith of Dante.
Dante is then blessed and encircled by *l'appostolico lume*, the apostolic light.

This experience of 'at-one-ment' that Dante describes could be taken as the Christian 'enlightenment', paralleling to a degree the Buddhist *Nirvana*. It is significant that Hui-Kuo (the Tang master of esoteric Buddhism) is recorded as saying to his famous student Kukai:

> Since the esoteric Buddhist teachings are so profound
> as to defy expression in writing, they are revealed
> through the medium of images to those who are yet to be
> enlightened . . . The sight of these may well enable one to
> attain Buddhahood.[72]

Dante, in wishing to share his vision with us, has to brave his own realization of the limits of words. Yet he persists within the vision of the Rose with its multiple sacred 'petal people' until he himself is taken into the cosmic vortex of the whole, which is:

> The Love that moves the sun and the other stars.
> (*L'amor che move il sole e l'altre stelle.*)

The 'garden' and the shining white Rose take up the most sacred figures of the Christian theogony in a blissful spiral of overwhelming geometric light. Beatrice, who accompanies Dante in this vision of Paradise, explains the geometry of light Dante is experiencing:

> . . . from the point of light
> Dependeth Heaven, and all things that exist.

> Look on that circle most conjoined with it;
> And know, its motion is so swift by aid
> Of love, whose kindling spurs its onward flight.

Dante describes his own state of mind during this vision:

> Thus did my mind in the suspense of thought
> Gaze fixedly, all immovable and intent,
> And ever fresh fire from its gazing caught.

This caused Dante to lament:

> To my conception O how frail and few
> My words!

The correspondence between quite different spiritual traditions cannot but alert us to an underlying universality. Both traditions here — the Buddhist and the Christian — use very similar symbolic language: a 'garden', a 'flower', 'light', 'love', a radiant single 'point' and the geometry of pure light.

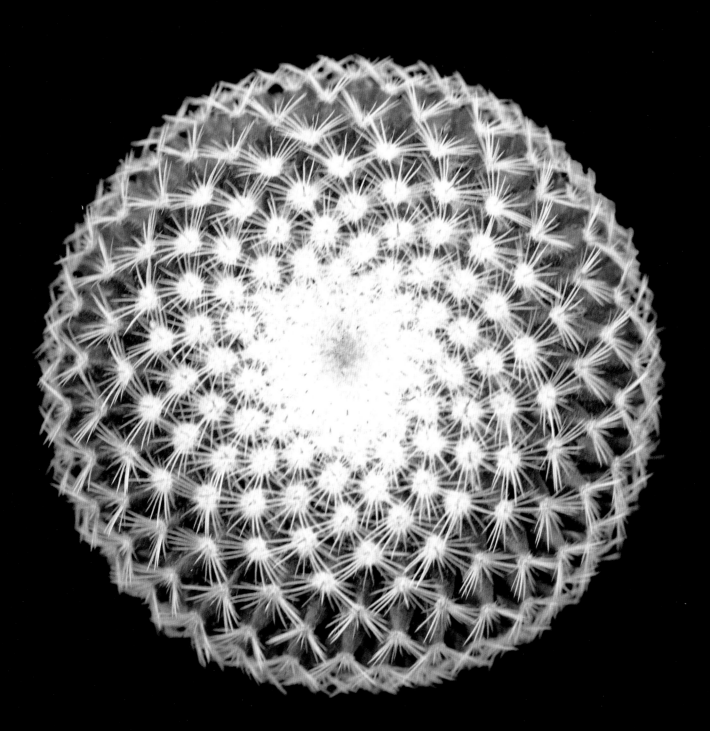

The final perfection of the human intellect is to be united to the angels by contemplation.

St Thomas Aquinas

Lysimachia Nummularia

It is called moneywort
for its "coinlike" leaves
and perhaps its golden flowers.
I love it because it is
a naturalized exotic
that does no harm,
and for its lowly thriving,
and for its actual
unlikeness to money.

Wendell Berry

Appendix

Coming to Terms with the Latin Names of Flowers

If you want to know about Bamboo — go to Bamboo

Basho

With thanks to Capel Manor College for the information on the Latin names of flowers. The comments, however, are our own.

Flowers, rather like each part of our bodies, have familiar as well as internationally recognized Latin names.

Most flowers have their local and national names, usually referred to as the 'common names'. These are recommended to be of primary value as they, more often than not, hold symbolic stories and tales that have deeper meanings. However, knowing the Latin names naturally enables one to communicate worldwide with more accuracy.

Both the Latin and the local names for flowers provide ways for getting to know their nature and significance; both to ourselves poetically, one might say, and to the human family as a whole. This means recognizing the conventional meaning of a flower as well as its symbolic or mythological meaning.

Both local and symbolic meanings, as well as the Latin names, hold disadvantages too. One may feel trapped by the 'common' name with its conventional symbolic meaning — as well as be trapped by the Latin names thinking that because you know them you therefore need search for meaning no further.

However we offer what we hope will be a worthy introduction to the binomial Latin naming system for flowers — devised by that outstanding Swedish botanist, Carl Linnaeus (1707–1778). Although this looks relatively recent in comparison to how long humans must have been naming flowers and plants, what Linnaeus did was to gather conventions originating among the Greeks (remember that Aristotle was a botany student of Plato's for at least eighteen years) which continued into Roman times and further into the Latin of the Christian Monasteries of the Middle Ages. Latin is a 'complete' language that has ceased to be developed and changed and therefore it has a 'timeless' value. This is where and why Latin was adopted by Linnaeus.

In two major works Linnaeus published his system of classification: *Genera Plantarum* and *Species Plantarum*. Linnaeus's contribution was first to group plants into families (for example *Rosaceae*) then give each individual plant two names, (hence the 'binomial' system).

Put simply, the first part of the name, the *genus*, indicates how the plant is most closely related to other plants (for example *Ilex*, holly) and the second is the *species* name (for example, *aquifolium*). Putting together the two names (*Ilex aquifolium*) provides a complete name by which the plant in question is known. By this method, Linnaeus devised a method of giving each plant a unique name. Any 'man-made' cultivars or hybrids constitute the third part of the name.

The genus and the species are always written in italics and the genus always begins with a capital letter. The cultivar or hybrid is in plain text, has capital letters or is within inverted commas, for example *Ilex aquifolium* 'Silver Queen'.

This Linnaean system was ratified internationally in both 1980 and 1988, finally being called the 'International Code of Botanical Nomenclature'. Thus worldwide we all have an agreed code which satisfies scientific, horticultural and commercial interests.

It is doubtful if poetic values have been claimed for dear Linnaeus's system, but this clearly was not its purpose.

So currently we have a fourfold classification system in Latin:

First: *Plant Families*

This is the basic unit of difference for all higher plants. A definition of 'family' is 'all descendants of a common ancestor and distinguished by common features'. In current botany this means grouping particular plants based on the botanical structure of their flowers, irrespective of external characteristics such as height, colour, foliage, etc. Examples include:

Rosaceae = Rose family
Ranunculaceae = Buttercup family
Liliaceae = Lily family

Within one family a huge variety of very diverse plants may occur. For instance *Rosaceae* includes trees (e.g. Hawthorn / *Crataegus*) shrubs (Mock Orange / *Philadelphus*), garden plants (Lady's Mantle / *Alchemilla*) and fruits (Strawberries / *Fragaria*).

Second: *Genus* (plural: *genera*)

This is a subdivision within a family. Linnaeus specifically chose a plant that he considered would best typify the genus he had in mind. This became his model. We can say that *genera* are based on many plant characteristics.

Within the family *Rosaceae* we find these *genera*:

Malus	*Prunus*	*Crataegus*	*Rosa*	*Sorbus*
Apples	Plums	Hawthorn	Roses	Mountain Ash
	Cherries			Whitebeam

A family may have only one genus or it may have many. The important Daisy family, called *Compositae* by Linnaeus, has over one thousand genera. The separate genera within one family may be closely related or quite different.

Third: *Species*

This identifies individual plants within a genus and has a name which is solely its own. For example, *Rosa canina*, commonly known as Dog Rose.

The 'species' name often indicates a specific plant characteristic, for instance *hirsuta* (hairy), *alba* (white) and *quinquefolia* (with foliage made up of five leaflets or leaves). It may also be geographical, like *chinensis* (Chinese) and *japonica* (Japanese). And there are many examples where the plant breeder or finder's name is used, as with *Magnolia wilsonii*, discovered in 1904 by Ernest Henry Wilson.

Fourth: *Cultivar* (written in English since 1959)

This name raises many issues although it is based on the simple meaning of cultivation. Culture is a close associate of 'to cultivate', and both are the basis of what eventually became civilization. This latter term, culture, relates to a society having successfully cultivated and stored sufficient foodstuffs (mainly grain) to support a diverse number of human activities and mutually beneficial skills.

So cultivars are basically *variants* that have been arrived at by human attention, cultivation, and the desire to 'improve' on the natural species through making them more disease-resistant for example.

We use the word 'wild' as a term which tends to indicate uncultivated species but also as a form of 'original' and even 'normal' specimens of plant life.

Cultivars can be categorized as plants that have received careful and patient breeding over the centuries. But they also come under the umbrella of plants that have arisen through mutation. This principle of mutation causes a sudden change in the genetic material of a cell, which may — and often does — cause it and all the cells derived from it to differ in appearance and behaviour from the original norm. Although mutation is cited as valuable in the history of a species, man-induced mutations are far from valuable in many instances. Cultivars arrived at by chemical or radiation action are what has been called 'genetic engineering'. This is politely called 'recombinant DNA technology'.

The driving force behind such activity is usually claimed to be 'beneficial', but the meaning of 'beneficial' and the long term results have often been questioned. Too often it is commercially driven by large seed-selling manufacturers and in many instances has been a disaster to small farmers who discovered the crops they harvested from such seeds *did not have fertile seeds* themselves!

It is important however to cite not just the negative outcome of certain kinds of cultivars and cultivation itself. Cultivars have been achieved throughout the history of human agricultural activity. This is particularly apparent with normal foodstuffs such as the grains and a good example is the common pea vegetable (*Pisum sativum, DC leguminosae*). Although this annual plant is of uncertain origin it is believed to be originally indigenous to central Europe or the mountainous region of Western Asia.

This beautifully simple plant with its 'packaged' green spherical contents has a different name in French, German, Finnish, Dutch, Danish, Italian, Spanish and Portuguese. This alone justifies the use of the Latin name for universality.

Most of us are also aware of the beautifully delicate perfume that the 'Sweet-pea' flower, a close relative of the vegetable, emits.

In *The Vegetable Garden* by Vilmorin-Andrieux,[73] first published in 1885 with exceptionally beautiful hand-drawn and engraved illustrations, there are twenty-eight normal-sized English named varieties of Peas, eight dwarf varieties, nine tall climbing varieties (wrinkled), and eight half-dwarf kinds. Next, there are seven varieties of 'sugar peas', all clear and well-illustrated at actual pod size. Then there are twenty-five other 'English' varieties, followed by thirty more varieties of wrinkled peas. After this we are given twenty-seven more variations of shelling peas called 'French'. These are followed by ten kinds of French 'Sugar-peas'. Next we find there are nine varieties of German-peas. Even further there are a few varieties of what are called 'Grey-peas' which are mostly fed to cattle. This remarkable book gives examples of 205 varieties from only three European nations. Surely a good example of tireless variety in cultivation even before patenting or commercial branding came in!

The same remarkable book gives eleven kinds of 'normal' Strawberries and forty-two hybrid Strawberry varieties. Knowing how delicious the Strawberry fruit is this is surely not surprising. Monsieur Vilmorin and his collaborators confess that the numbers of variations are so great that they have only selected fifty-three to mention specifically. The author also excuses himself for including a fruit because it is 'generally cultivated with kitchen garden crops'.

Finally, there are fifty kinds of common cabbage, sixteen Savoy cabbage varieties and forty-eight kinds of garden beet. These include the types fed to cattle.

Returning to flowers and their 'breeding', we might ask if it is valid to regard the flower in question as having any response to what it is being subjected to. If Sir J. C. Bose, whom we discussed earlier, established precisely over his lifetime that the plant world certainly *does* respond to human thought (and presumably feelings) then it would seem to follow that each chemically modified plant might well have its own response in recent history. We have certainly had examples of proposed and even actual human eugenics practised under the argument of improving the quality of the human population, by 'selective breeding' on the one hand and 'elimination of characteristics' on the other. The values and judgment exercised in such programmes have aroused particular scrutiny and reaction. Maybe we must soon seek similar 'rights' and 'protection' for the plant world.

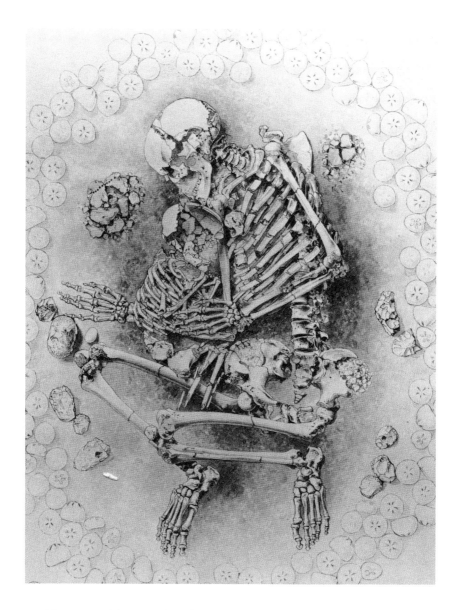

The very careful surrounding of the body in this neolithic grave shows a special recognition of the fossilized fivefold sea creatures. This indicates a sensitivity to symmetry that is hardly likely to have excluded flowers.

Endnotes

1. Corbin, H. *The Man of Light in Iranian Sufis.*

2. Plato, *Timaeus,* Volume VII, p. 61.

3. Plato, *Timaeus,* p. 53.

4. Johann Wolfgang von Goethe (1740–1832). German poet, painter, scientist, luminary and contemporary of J. M. W. Turner. His Strasbourg teacher Herder introduced him to Gothic architecture and to the meaning and beauty of the natural world. Also an enthusiast for Shakespeare. While in Weimar in the 1770s, Goethe learned about agriculture, horticulture and even mining as the main support for welfare of that duchy — which led to his preoccupation with the natural sciences. These took an increasing amount of his time as he matured. He made an important visit to Italy in 1786–1788 which he considered the 'climax' to his life. He was initiated into Neo-Classicism, the content of which he never divulged but his literary works such as *Iphigenie auf Tauris* (1787) and the drama *Torquato Tasso* (1790) contain the ideas and inspiration he received in Rome. Goethe as a European led a powerfully influential life.

5. See Gardner, H. *Multiple Intelligences: The Theory in Practice.*

6. Schumacher, E. F. *Small Is Beautiful.*

7. Sheldrake, Rupert, *A New Science of Life.*

8. Plato, *Timaeus,* 29-D.

9. Iamblichus from *The Life of Pythagoras*: '. . . that first essence is the nature of Number and "reasons" (*Logoi* or productive principle) which pervades everything'. *The Golden Chain* by Algis Uždavinys, page 19.

10. The 'four' or quadrivium as they later were called by the Latins are the arts/sciences; the first two are Arithmetic and Geometry, the other two are Harmony and Astronomy (or Cosmology).

11. Coomaraswami, A. K. *Time and Eternity.*

12. See Critchlow, K. *Order in Space.*

13. Plato, *Republic*, II 526c-527c.

14. Chaturvedi, Dr Sharda, *The Scientific Attitude of Vedas*, www.gavaksha.org.

15. Naydler, J. *The Future of the Ancient World: Essays on the History of Consciousness,* Part One 'The Heart of the Lily'.

16. Plato, *Republic,* VI 509

17. Krishnamurti, J. *Krishnamurti on Nature and the Environment.*

18. See Seligman, M. *Authentic Happiness.*

19. Haviland-Jones J., *et al.* (2005) 'An Environmental Approach to Positive Emotion: Flowers' in *Evolutionary Psychology* Volume 3: 104-132.

20. Plato, *Timaeus,* 29-C. Timeaus says, 'Enough if we adduce probabilities as likely as any others; for we must remember that I who am the speaker and you who are the judges, are only mortal, and we ought to accept the tale that is most likely'.

21. Mees, G. H. *The Book of Signs* (also *of Battles* and *of Stars*).

22. See Cowen, Painton *The Rose Window.*

23. Cleene, M. de and Lejeune, M. C. *Compendium of Symbolic and Ritual Plants in Europe.*

24. Plato, *The Collected Dialogues.*

25. Proclus also confirms the intrinsic Platonic fourness in *The Elements of Theology.* Proposition 20 states: 'Beyond all bodies is the soul's essence; beyond all souls, the intellective principle; and beyond all intellective substances, the One'.

26. Plato, *Timaeus,* p. 16.

27. See Hopper, Vincent Foster *Medieval Number Symbolism.* This book is highly recommended.

28. See Halevi, Z'ev ben Shimon *A Kabbalistic Universe.*

29. Schimmel, Annemarie *The Mystery of Numbers.*

30. Bunning, E. *Die Physiologische Uhr.*

31. Brown F. A. 'Why is so little known about the biological clock?' in Halbert, Scheving & Pauly, *Chronobiology,* pp 689ff.

32. Critchlow, Keith, *Order in Space,* p. 4.

33. Plato, *Timaeus,* p. 51.

34. Plotinus, *Ennead,* V 8th Tractate.

35. Plato, *Republic,* 527A-C.

36. HRH The Prince of Wales; Juniper, Tony; Skelly, Ian, *Harmony: A New Way of Looking at the World.*

37. Exod. 34:30, 34:35; 2Cor. 3:13; Matt. 17:2; Rev 10:1.

38. See Berry, Wendell, *Life Is a Miracle.*

39. See Olsen, Scott, *The Golden Section, Nature's Greatest Secret.*

40. Ibn Gabirol, *Selected Poems of Solomon Ibn Gabirol.*

41. See Sheldrake, *A New Science of Life.*

42. See Plato, *Timaeus,* 27D-28. Also stated by Socrates in the *Republic,* see the 'Divided Line', chapter xxiv VI 509.

43. Plato, *Timaeus,* 29A and 34C.

44. Plato, *Republic,* 509 D-511E.

45. It is important here to state our conviction that the higher purpose of geometry is to 'participate', body, soul and spirit, in the objective universal laws that govern and cohere our universe. The activity can lead us directly to the centre of our own understanding which unifies us with the whole. *It is not an exercise in mechanics.*

46. See Needham, Joseph, *Science and Civilization in China.*

47. Matt. 6:28.

48. Nasr, S. H. *Knowledge and the Sacred,* p. 215.

49. It is in the very word 'science' that the most important changes in perception have to be made. Dr S. H. Nasr, the contemporary Muslim master of philosophy, has written with great clarity on this issue of the meaning of the word science, in *Science and Civilization in Islam.*

50. From Hardin, Garret, *Biology: Its Principles and Implications.*

51. Rupert Sheldrake's works *(A New Science of Life)* and contributions are a good example of the challenge that is currently being debated. Also to be recommended is the work of Brian Goodwin, *How the Leopard Changed Its Spots: The Evolution of Complexity.*

52. See Burckhardt, Titus, *Alchemy.*

53. See Jenny, Hans, *Cymatics: A Study of Wave Phenomena,* and Lauterwasser, Alexander, *Water Sound Images: The Creative Music of the Universe.*

54. Rumi, Jeláluddin, *The Ruins of the Heart.*

55. Traas, Jan & Vernoux, Teva, 'The shoot apical meristem: the dynamics of a stable structure'. *Philos Trans R Soc Lond B Biol Sci.* June 29; 357 (1422): 737–47.

56. See Martineau, John, *A Little Book of Coincidence*, pp. 32 & 47.

57. Bell, Adrian, *Plant Form*, pp. 219–21.

58. Chu Hsi, *Chu Tzu Yu-lei (Analects of Master Chu)*, 95:239. See Wing-tsit Chan et al. *Reflections on Things at Hand.*

59. Chu Hsi, *Chu Tzu Yu-lei (Analects of Master Chu)*, 95:23a

60. From Krishnamurti J. *Krishnamurti on Nature and the Environment.*

61. St Matthew is quite clear on the tradition which reports Jesus on the meaning of the field. 'The field is the world and the good sons mean the sons of the kingdom, the tares or weeds are the sons of the evil one'. See Matt. 13:39.

62. Halevi, Z'ev ben Shimon, *The Kabbalist at Work.*

63. Mees, G. H., *The Book of Signs,* p. 180.

64. For an important piece of contemporary wisdom see Berry, Wendell, *Life is a Miracle.*

65. Translation from *The Mathnavi*, adapted slightly by the author.

66. Margulis, L., and Sagan, D. *Microcosmos: Four Billion Years of Microbial Evolution*, p. 29.

67. Sheldrake, Rupert, *A New Science of Life.*

68. This ladder is to be found in all authentic philosophical traditions developed by mankind as far as this author's experience and research shows. The work of Z'ev ben Shimon Halevi, *The Kabbalist at Work,* on the Kabbalistic tradition is one such valuable guiding light in this area.

69. See Plato, *Republic* op. cit., VI 509-511B.

70. Critchlow, K., *Order in Space,* p. 5.

71. All quotations from the *Paradiso* are from Laurence Binyon's translation first published in 1943. Available in *The Portable Dante.*

72. From Birnbaum, Raoul, *The Healing Buddha*, p. 77.

73. Vilmorin-Andrieux, *The Vegetable Garden.*

Bibliography

Bell, Adrian D. (1991) *Plant Form,* Oxford University Press, UK; revised ed. 2008, Timber Press, Portland, Oregon, and London, UK.

Berry, Wendell (2001) *Life is a Miracle,* Counterpoint Press, Berkeley.

—, (2005) *Given: New Poems,* Shoemaker & Hoard, Berkeley.

Birnbaum, Raoul (2003) *The Healing Buddha*, Shambhala Publications, Boston.

Bunning, E. (1963/1977) *Die Physiologische Uhr,* 2nd and 3rd eds., Springer, Berlin.

Burckhardt, Titus (1967) *Alchemy*, Stuart and Watkins, London.

Cleene, M. de & Lejeune, M.C. (2003) *Compendium of Symbolic and Ritual Plants in Europe,* two vols., Mens-en-Cultuur Uitgevers, Ghent, Belgium.

Coomaraswamy, Ananda (1947) *Time and Eternity,* Artibus Asiae, Ascona, Switzerland.

Corbin, H. (1994) *The Man of Light in Iranian Sufism,* trans. Nancy Pearson. Omega Publications, New York.

Cowen, Painton (2005) *The Rose Window*, Thames & Hudson, London.

Critchlow, K. (2008) *On Finding One's Marbles,* Kairos Publications, Devon.

— (1969/2000) *Order in Space,* Thames & Hudson, London.

—, (1979/2007) *Time Stands Still,* Floris Books, Edinburgh.

Dante, *The Portable Dante* (1995) Penguin Books.

Eliade, Mircea (1996) *Patterns in Comparative Religion,* trans. Rosemary Sheed, University of Nebraska Press.

—, (1959/1987) *The Sacred and The Profane*, trans. Willard Trask, Harcourt, Florida.

Gabirol, Solomon Ibn, *Selected Poems of Solomon Ibn Gabirol*, translated by Peter Cole (2001), Princeton University Press, Princeton, USA, and Woodstock, UK.

Gardner, H. (1993) *Multiple Intelligences: The Theory in Practice,* Basic Books, New York.

Goodwin, Brian (2001) *How the Leopard Changed Its Spots: The Evolution of Complexity*, Princeton University Press.

— (2007) *Nature's Due,* Floris Books, Edinburgh, UK.

Halberg, F., Scheving, L.E. & Pauly, J.E. (eds.), (1974) *Chronobiology,* Igaku Shoin, Tokyo.

Halevi, Z'ev ben Shimon (2008) *A Kabbalistic Universe*, The Kabbalah Society, London.

Hardin, Garrett & Bajema, Carl (1978) *Biology: Its Principles and Applications,* W.H. Freeman & Co., San Francisco and London.

Hopper, Vincent Foster (1938/2000) *Medieval Number Symbolism*, Dover Publications.

Jenny, Hans (2001) *Cymatics,* Macromedia Press, New Hampshire.

Krishnamurti, J. (1992) *Krishnamurti on Nature and the Environment,* Orion Books, London.

Lauterwasser, Alexander (2007) *Water Sound Images: The Creative Music of the Universe*, Macromedia Press, New Hampshire.

Lings, Martin (2006) *Symbol and Archetype,* Fons Vitae, Louisville.

Margulis, L., and Sagan, D. (1997) *Microcosmos: Four Billion Years of Microbial Evolution,* p. 29. University of California Press, Berkeley.

Martineau, John (1995) *A Book of Coincidence,* reissued 2001 as *A Little Book of Coincidence,* Wooden Books, Glastonbury; Walker & Co. New York.

Mees, G. H. (1953) *The Book of Signs,* Kluwer, Deventer, Holland (also *The Book of Battles* and *The Book of Stars).*

Nasr, S.H. (1984) *Knowledge and the Sacred,* Edinburgh University Press.

— (1968, 1987) *Science and Civilization in Islam,* Harvard University Press; Islamic Texts Society, Cambridge UK.

Naydler, J. (2009) *The Future of the Ancient World: Essays on the History of Consciousness,* Abzu Press, Oxford.

Needham, Joseph (1954–58) *The Science and Civilization of China,* seven volumes, Cambridge University Press. A 'shorter' series has also been published by CUP.

Olsen, Scott (2006) *The Golden Section: Nature's Greatest Secret,* Wooden Books, Glastonbury; Walker and Co., New York.

Plato (1961) *The Collected Dialogues,* ed. Hamilton and Cairns, Bollingen Series, Princeton University Press.

—, (1952) *Timaeus,* translated by R. G. Bury, Loeb Classical Library, Harvard University Press.

—, *Republic,* translated by Robin Waterfield (1993-4) World's Classics, Oxford University Press.

Plotinus (1989) *The Enneads,* Loeb Classical Library, Harvard University Press.

Proclus (1994) *The Elements of Theology,* translated by E. R. Dodds). Oxford University Press.

Querido, René (1990/2008) *The Golden Age of Chartres,* Floris Books, Edinburgh.

Rumi, Jelaluddin (2002) *The Mathnawi,* trans. E. H. Whinfield. Duncan Baird, London.

—, (1981) *The Ruins of the Heart,* trans. Edmund Helminski. Windrush, Witney, Oxfordshire.

Schimmel, Annemarie (1994) *The Mystery of Numbers,* Oxford University Press, New York.

Schumacher, E. F. (1973) *Small Is Beautiful,* Vintage, London, new edition 1993.

Seligman, M. (2003) *Authentic Happiness,* Nicholas Brealey, London.

Sheldrake, R. (2009) *A New Science of Life,* 3rd ed., Icon Books, London.

Skelly, Ian (2010) *Harmony: A New Way of Looking at the World,* Blue Door, HarperCollins, London.

Uždavinys, Algis (2004) *The Golden Chain: An Anthology of Pythagorean and Platonic Philosophy,* World Wisdom Books, Bloomington, Indiana.

—, (2008) *Philosophy as a Rite of Rebirth: From Ancient Egypt to Neoplatonism,* Prometheus Trust, Westbury.

Vilmorin-Andrieux, M. M. (1977) *The Vegetable Garden* (facsimile of 1885 edition), John Murray, London.

Wing-tsit Chan, *et al.* (1967) *Reflections on Things at Hand,* Columbia University Press, New York.

Image Credits

Abe Collection, Municipal Museum, Osaka: 100

Alinari Archives,Florence: 64, 65, 101

The Art Institute of Chicago: 398

Julian Barnard: 33, 135, 162, 394

The Bridgeman Art Library / Giraudon / Bibliotheque Municipale, Reims: 39

The Bridgeman Art Library / The Stapleton Collection / Österreichische
    Nationalbibliothek, Vienna: 91

Bridgeman Art Library, London / SuperStock: 103

British Museum: 104, 111

Graham Challifour: 81, 387

CORBIS © Burstein Collection: 87

Faber and Faber: 38

Flower Remedy Programme: 41

Getty Museum: 87

Ana-Maria Giraldo: 57, 89, 296

Ben Shimon Halevi: 119

Alexander Lauterwasser: 373

Andrew Lawson: 58

Linnaeus Society: 72

Robin Heath: 76, 201

John Martineau: 77, 78, 380

Charles Deering McCormick Library of Special Collection, Evanston, IL: 67

Trish Mulholland: 264

NASA: 28, 55, 74

Oxford University Press: 39

Philipp Allard.© 2011. Photo Scala, Florence / BPK, Bildagentur für Kunst, Kultur und
    Geschichte, Berlin: 423

Joerg P. Anders.© 2011. Photo Scala, Florence / BPK, Bildagentur für Kunst, Kultur und
    Geschichte, Berlin: 422

Photo Scala, Florence / BPK, Bildagentur für Kunst, Kultur und Geschichte, Berlin: 423

Science Photo Library: 82, 97, 143, 375

3D4Medical.com / Science Photo Library: 144

Farmilab / Science Photo Library: 154

Manfred Kage / Science Photo Library: 366

David Scharf / Science Photo Library: 54

Dr Richard Kessel and Dr Gene Shih, Visuals Unlimited / Science Photo Library: 372

Archie Young / Science Photo Library: 63

Hans Silvester / Rapho / Gamma Camera Press, London: 90

Babak Tafreshi: 170

Victoria & Albert Museum: 98

Index

Floris Books

For news on all our **latest books**,
and to receive **exclusive discounts**,
join our mailing list at:

florisbooks.co.uk

Plus subscribers get a FREE book
with every online order!

We will never pass your details to anyone else.